Study Guide

Sandra Bobick
Community College of Allegheny County

Laurel Roberts
University of Pittsburgh

BIOLOGY

A GUIDE TO THE NATURAL WORLD

FOURTH EDITION

DAVID KROGH

PEARSON

Benjamin
Cummings

San Francisco   Boston   New York
Cape Town   Hong Kong   London   Madrid   Mexico City
Montreal   Munich   Paris   Singapore   Sydney   Tokyo   Toronto

Acquisitions Editor: *Star MacKenzie*
Supplement Project Editor: *Blythe Robbins*
Editorial Assistant: *Erin Mann*
Senior Marketing Manager: *Jay Jenkins*
Managing Editor: *Mike Early*
Senior Production Supervisor: *Shannon Tozier*
Production Editor: *Leslie Austin*
Text Design: *Pre-PressPMG with S4 Carlisle Publishing Services*
Composition: *Pre-PressPMG*
Illustrations: *Pre-PressPMG with Imagineering Media Services, Inc.*
Cover Design: *Riezebos Holzbaur Design Group*
Cover Production: *Seventeenth Street Studios, Inc.*
Manufacturing Buyer: *Jeffrey Sargent*
Cover and Text Printer: *Malloy, Inc.*
Cover Photo Credit: *shell © Andrew Mounter/Taxi; frogs © Gail Shumway/Taxi; fungus © Theo Allofs/Photonica*

ISBN 10-digit 0-13-225478-6
     13-digit 978-0-13-225478-6

PEARSON

Benjamin
Cummings

1 2 3 4 5 6 7 8 9 10—MAL—12 11 10 09 08
www.pearsonhighered.com

CONTENTS

TO THE STUDENT

Welcome to the *Study Guide* to accompany *Biology: A Guide to the Natural World, Fourth Edition*. This supplement to your textbook has been designed to make you as successful as possible in your study of biology. The *Study Guide* is one of several resources available to you. It is not meant to replace your lecture notes or your textbook, but rather to reinforce and test your understanding of the concepts and details presented in class.

The *Study Guide* is divided into 36 chapters that correspond directly to those in the textbook. Within each chapter you will find the following:

- **Basic Chapter Concepts:** an outline of the broad concepts covered in the chapter.

- **Chapter Summary:** an outline of the major topics broken down section by section.

- **Word Roots:** definitions of word roots found in the chapter's key terms.

- **Key Terms:** a comprehensive list of important terms from the text, including blanks where you can fill in their definitions.

- **Flash Cards:** cards featuring each chapter's key terms and definitions, which you can tear out and use to prepare for exams.

- **Self Test:** compare-and-contrast, short-answer, and multiple-choice questions that test your understanding of the chapter's major concepts.

- **What's It All About?:** a sample essay question appearing at the end of each chapter to help you practice organizing information so that you can write a clear, concise answer supported by specific examples. In the first 10 chapters, we'll outline an approach to answering these "big picture" questions, but beginning with Chapter 11, you're on your own.

- **Answer Key:** answers to the Self Test questions for each chapter, allowing you to gauge your understanding of the material.

The *Study Guide* not only will help you review important concepts, but also will provide an opportunity to explore the biology in your life and in the world around you. For additional study resources, go to www.mybiology.com.

CHAPTER 1 SCIENCE AS A WAY OF LEARNING: A GUIDE TO THE NATURAL WORLD

Basic Chapter Concepts

* Science is both a body of knowledge and a method to gain that knowledge.
* Science and technology play an important role in the lives of Americans.
* Science is a process that is carried out through observation and testing of hypotheses.
* Biology is the study of living things.

CHAPTER SUMMARY

1.1   How Does Science Impact the Living World?

* American citizens turn to science for answers concerning their health, the environment, and ethical decisions related to new advances in medicine.
* Most Americans have an uneven knowledge about science.

1.2   What Is Science?

* Scientific evidence consists of observations and the results of experiments done to account for the observations.
* Theories are statements of general principles, supported by evidence, about the natural world.
* Experiments proceed using the scientific method, a process that involves testing the validity of a proposed explanation (hypothesis) in a controlled manner.
* No hypothesis or theory is ever proven absolutely in science, but if we fail to disprove theories often enough, we believe them to be true.

1.3   The Nature of Biology

* Biology is the study of highly organized, living things.
* Living things can be differentiated from nonliving things by a suite of characteristics—there is no single factor that differentiates living from nonliving.
* Living things are highly organized, and this organization occurs in a hierarchy such that relatively simple and small components become integrated to make larger, more complex components.

1.4   Special Qualities of Biology

* Before the nineteenth century, biology was largely a descriptive science dealing with listing and cataloguing living things.
* Evolution is the chief underlying principle of biology.
* Biology is distinct from a study of the physical sciences, such as physics, because it deals with living organisms rather than universally applicable rules.

WORD ROOTS

bio- = life (e.g., *bio*logy is the study of living things)

eco- = home (e.g., *eco*logy is the study of how organisms interact with one another and the environment)

homo- = same (e.g., *homo*stasis is the regulation and maintenance of the internal conditions of an organism)

mol- = molecule (e.g., *mol*ecular biology is the study of molecules found in living organisms)

KEY TERMS

biology _____

ecology _____

evolution _____

homeostasis _____

hypothesis _____

life sciences _____

molecular biology _____

organismal biology _____

physical sciences _____

physiology _____

science _____

scientific method _____

theory _____

variable _____

FLASH CARDS

To use the flash cards, tear the page from the book and cut along the dashed lines. The key term appears on one side of the flash card, and its definition appears on the opposite side.

biology	organismal biology
ecology	physical sciences
evolution	physiology
homeostasis	science
hypothesis	scientific method
life sciences	theory
molecular biology	variable

the study of whole organisms	the study of life
the natural sciences not concerned with life	the study of the interactions that living things have with each other and with their environment
the study of the physical functioning of animals and plants	any genetically based phenotypic change in a population of organisms over successive generations; evolution can also be thought of as the process by which species of living things can undergo modification over successive generations, with such modification sometimes resulting in the formation of new species
a means of coming to understand the natural world through observation and the testing of hypotheses; also, a collection of insights about nature, the evidence for which is an array of facts	the maintenance of a relatively stable internal environment in living things
the process by which scientists investigate the natural world; the scientific method involves the testing of hypotheses through observation and experiment, as aided by the tools of statistics	a tentative, testable explanation of an observed phenomenon
a general set of principles, supported by evidence that explains some aspect of nature	a set of disciplines that focus on various aspects of the living world; the life sciences include biology and related disciplines such as medicine and forestry
an element of an experiment that is changed in comparison to an initial condition	the investigation of life at the level of its individual molecules

SELF TEST

Once you have finished studying this chapter, close your books, grab a pencil, and spend the next 15 to 20 minutes completing this practice test.

Compare and Contrast

For each of the following paired terms, use a sentence of comparison ("Both") and a sentence of contrast ("However,").

variable/constant
theory/hypothesis
atoms/molecules
cell/organisms
population/community

Short Answer

1. Imagine you are a citizen of the world prior to 1850. Why do you think people believed that life could be created spontaneously?

2. What is the relationship between a hypothesis and a theory?

3. Why must a hypothesis be falsifiable?

4. Why is it necessary to define life by a list of criteria instead of a single attribute?

Multiple Choice

Circle the letter that best answers the question.

1. What is the possible explanation for a series of observations?
   a. guess
   b. theory
   c. law
   d. hypothesis
   e. hunch

2. _____ is the unifying principle of biology.
   a. Molecular biology
   b. Evolution
   c. Physics
   d. Physiology
   e. Homeostasis

3. Which of the following represents a series ranked from smallest to largest?
   a. population → community → biosphere
   b. atom → molecule → cell
   c. tissue → organ → organism
   d. cell → organelle → organism
   e. a, b, and c

4. Which of the following is/are shared by all living organisms?
   a. All have the ability to assimilate energy.
   b. All have the ability to transmit genetic information.

   c. All are made up of building blocks called organs.
   d. all of the above
   e. a and b only

5. Your spaceship crashes on Planet 2X-Alpha. While waiting for help, you try to discover as much as possible about your temporary home. To uncover facts about your new "natural world," you would first:
   a. design theories to fit facts into.
   b. do experiments.
   c. draw conclusions.
   d. make observations.
   e. write up your results.

6. What's the second thing you should do to uncover facts about Planet 2X-Alpha?
   a. design theories to fit facts into
   b. do experiments
   c. draw conclusions
   d. make observations
   e. write down what others have observed about the planet

7. Which of the following is an example of a testable (falsifiable) hypothesis?
   a. Plants grow best on days when the temperature rises above 40 degrees.
   b. The best flowers are pink.
   c. Apples are nice.
   d. Many things happen; not all can be explained, but some can be experienced.
   e. Students with glasses are more intelligent than those who wear contacts.

8. Your answer for the previous question was the best choice because:
   a. it had the most words.
   b. it had numbers.
   c. it used inductive reasoning to generate universal laws.
   d. it was most likely to generate a controlled test.
   e. it relied entirely on personal opinion.

9. Which of the following is the simplest thing that shows the characteristics of life?
   a. a population of mice
   b. a DNA molecule in the brain of a mouse
   c. a mouse
   d. the brain of a mouse
   e. a mouse brain cell

10. Pasteur carried out his experiment to demonstrate that:
   a. the Greeks were wrong with their theories.
   b. it is actually airborne organisms that give rise to other living things.
   c. people shouldn't leave meat out at room temperature.

d. life could arise spontaneously.

e. the scientific method always works.

11. Living organisms are different from inanimate objects because they:
    a. react to changes in their environment.
    b. are highly organized.
    c. must use energy to live.
    d. are composed of cells.
    e. all of the above

12. Which of the following levels of organization includes all of the others?
    a. cell
    b. molecule
    c. organism
    d. community
    e. biosphere

13. Humans usually have a body temperature of 37°C unless they are ill. This is an example of:
    a. metabolism.
    b. homeostasis.

c. biology.

d. adaptation.

e. all of the above

14. Scientific theory is based on:
    a. a higher authority.
    b. consensus of opinion.
    c. evidence.
    d. repeated experimentation.
    e. c and d

15. An ecologist in Brazil is studying how the nutrients and organisms on the bottom of the Amazon River affect bacterial growth in the river. This study would not involve which of the following levels of organization?
    a. population
    b. organism
    c. community
    d. ecosystem
    e. chemical

WHAT'S IT ALL ABOUT?

Here's a question to help you pull together what you've learned so far using this text.

Question: How does the fact that living organisms have a hierarchical organization impact the process of biological research?

1. **What is this question asking me to do?**

   Always stop first and figure out what the answer should look like. Is it a compare/contrast question? That means you need to provide some specific examples of similarities and differences. Does it ask you to defend a statement? Those questions require you to provide the experimental findings that support the claim made in the question. A more open-ended form of this type of question asks you to support or refute the statement. Again, you need to muster your factual evidence to defend your position. Does the question ask you to describe or document the effect of one thing upon another, as this question does? In this case, we need to collect the information that describes the theory and describe how each part affects some other process, in this case, biological research.

2. **Collect the evidence.**

   Our first step is to define, in detail, what is meant by a "hierarchical organization" and the "process" of biological research. We start by looking for information in the textbook.

Hierarchical Organization	Process of Research
Living things highly organized	Research starts with an observation that produces a question
Lower levels of organization integrated to create high ones (so higher levels don't exist without lower levels)	Questions lead to proposing possible assertions that can be experimentally tested and possibly found to be false (hypotheses)
Levels = atom, molecule, organelles, cells, tissues, organs, organism, population, community, biosphere	Theory = description of general principles that fits with all of the available factual evidence; has power to explain things but is also provisional
Levels define "building blocks"	

Collect all the information that you can even if you probably won't use all of it in your final answer.

3. Pull it together.

We are trying to understand the impact on the process of research, so let's start by asking how each part of the research process is affected by the hierarchy of life.

Starts with An Observation

Can make observations on any level of the hierarchy.

Observations made on a given level for any one organism should be the same or very similar to that found for that same level in other organisms. For example, the properties of a specific molecule found in a fish should be very similar to the properties of the same molecule when found in a mouse.

Questions Lead to Hypotheses

Hypotheses can be tested in different organisms as long as we consider the same level of organization—that is, we can test questions about a specific molecule by using molecules isolated from worms, mice, or people. Testing in different organisms and collecting reproducible results in each organism increases certainty in our findings.

Evidence Is Used to Produce a Theory

It's a little tricky to see how the hierarchical organizations can help us make general theories. It seems that if our research is limited to one level, the best we could do is to suggest theories about one level of organization.

Did any insights occur to you as you organized this information? What about the fact that the levels of the hierarchy are integrated? Does this imply that if we define some new property of the cell that we can then suggest a new property for the tissue? Likewise, does a newly identified cellular property imply that there must be new molecules (or old molecules with previously unidentified functions) involved? We can apply what we know about the hierarchy of life to the "problem" of research method and develop the new idea that research at any one level of the hierarchy can lead to new hypotheses about function at other levels of the hierarchy.

So Let's Write

The hierarchical organization of living organisms presents well-defined systems to observe and ask questions about. For example, if we are observing an organism eating, we can develop hypotheses about why it chooses that food (organism level), how easy or difficult it is to find that food (community level), how it digests that food (organ level), or how it extracts energy from that food (cellular level). So one observation can be explored on many different levels. Because all of these levels are integrated, our findings on any one level can lead us to make suggestions about function at higher or lower levels of organization. The result is that by collecting the results of research done at the molecular, cellular, tissue, organ, and community levels of organization, we can produce a general theory of how and why that organism chooses that particular food.

CHAPTER 2 FUNDAMENTAL BUILDING BLOCKS: CHEMISTRY, WATER, AND pH

Basic Chapter Concepts

- The fundamental building block of matter is the atom, an elemental particle composed of protons, neutrons, and electrons.
- The complex molecules of life are created from atoms through covalent, ionic, and hydrogen bonding.
- Chemical bonds define a molecule's physical, chemical, and biological properties.
- Water's unique structure and polar character facilitate its role as a biological solvent.
- We measure the dissociation of water, and other molecules, using the pH scale. Solutions with a low pH are acids; those with a high pH are bases.

CHAPTER SUMMARY

2.1 Chemistry's Building Block: The Atom

- The basic forms of matter are the elements, 92 pure substances that cannot be reduced to simpler components. A single elemental particle is called an atom.
- Atoms contain protons, neutrons, and electrons. Protons and neutrons occupy the nucleus of an atom and account for its mass. Electrons are essentially mass-less particles that occupy the space around the nucleus.
- Protons are positively charged and electrons are negatively charged. Atoms are electrically neutral, however, because the number of protons equals the number of electrons, ensuring no net charge.

2.2 Matter Is Transformed through Chemical Bonding

- We think of electrons surrounding the nucleus in layers, called shells, that reflect the energy content of the molecule. A limited number of electrons can occupy any given shell and, when a shell is filled, the atom exists in its most stable form.
- In covalent chemical bonding, the atoms involved share electrons to achieve a full outer valence shell. The total number of electrons does not change when a covalent bond is made. A molecule is created when covalent bonds join two or more elements.
- Electrons in a covalent bond may not be shared equally. Unequal sharing produces a small difference in the charge distribution around the molecule, so that one end is more positive and the other is more negative, creating a polarity to the bond. Polar molecules may associate through very weak bonds called hydrogen bonds.
- Ionic chemical bonding occurs when electrons are pulled from the shell of one atom into the shell of another, creating two charged atoms known as ions. The atom that has lost an electron becomes a positive ion; the one that gains an electron becomes a negative ion. The ions remain associated because their opposite charges attract each other.

2.3 Some Qualities of Chemical Compounds

- A molecule has a three-dimensional shape that determines its own chemical and biological activity and its ability to interact with other molecules.
- Shape and polarity determine how well molecules form homogeneous mixtures called solutions. Solvent molecules break the chemical bonds, usually ionic or hydrogen bonds, which hold solute molecules together.
- "Like" dissolves "like"—molecules of similar polarity can form solutions, such as water and salt, which are both polar molecules.

2.4   Water and Life
- Water works as a solvent because it can form hydrogen bonds with many molecules, thus causing them to dissolve into solution. It can also form a hydrogen bond with itself, which explains its ability to serve as a heat buffer, to become less dense in its solid state, and to have a high surface tension.
- Water cannot solubilize nonpolar molecules, such as hydrocarbons, because it cannot disrupt the attraction that holds one nonpolar molecule to another. Nonpolar molecules are referred to as hydrophobic because they do not interact with water, as hydrophilic molecules do.
- Water's ability to disrupt the hydrogen bonds formed when hydrophilic molecules interact creates acids and bases.

2.5   Acids and Bases Are Important to Life
- Acids and bases are substances that increase or decrease the concentration of hydrogen ions in an aqueous solution.
- The pH scale allows us to quantify how readily a molecule will lose a hydrogen ion (acidic) or gain a hydrogen ion (basic) when put in an aqueous solution.
- The pH scale is logarithmic—a solution at pH 5 has 100 times more hydrogen ions in it than a solution at pH 7. Smaller numbers mean that more hydrogen ions are present and the solution is more acidic; bigger numbers mean that the solution is less acidic and more basic.

WORD ROOTS

hydro- = hydrogen, water (e.g., a carbo*hydr*ate is made up of carbon, hydrogen, and oxygen)

ion = referring to the charged form of an atom (an *ion*ic bond forms between positively and negatively charged atoms, such as Na^+ and Cl^-)

alka- = basic (an *alka*line substance raises the pH when added to an aqueous solution)

polar = charge differential (the *polar*ity of water molecules causes them to have a positive and a negative pole)

KEY TERMS

acid *Any substance that yields hydrogen ions in solution; + than 7ph.*

acid rain

alkaline

atomic number

ball-and-stick model

base

buffering system

chemical bonding *electrons of 2 atoms interact and rearrange into new forms.*

covalent bond

electron

electronegativity

element

free radical

hydrocarbon

hydrogen bond

hydrophilic

hydrophobic

hydroxide ion

ion

ionic bonding

ionic compound _A compound composed of the unked ionic forms of 2 or more elements_

polarity _____

isotope _____

product _____

law of conservation of mass _____

proton _____

mass _____

reactant _A substance that goes into a chemical reaction; reactants intract to form the product._

molecular formula _____

specific heat _____

molecule _____

solute _The substance being dissolved by a solvent to form a solution._

neutron _____

solution _____

nonpolar covalent bond _____

solvent _____

nucleus _____

space-filling model _____

pH scale _____

structural formula _____

polar covalent bond _____

SELF TEST

Once you have finished studying this chapter, close your books, grab a pencil, and spend the next 15 to 20 minutes completing this practice test.

Compare and Contrast

For each of the following paired terms, use a sentence of comparison ("Both") and a sentence of contrast ("However,").

polar/nonpolar
inert/reactive
covalent/ionic
element/matter
solute/solvent

Short Answer

1. What is an element?

2. Which would have fewer electrons, an atom or its negatively charged ion?

3. Why is the three-dimensional shape of a molecule important?

4. Your home is in Boston, Massachusetts, and your college roommate is from Tucson, Arizona. Given that Boston is near the ocean and Tucson is not near any bodies of water, how would you expect daily and nightly temperatures to compare in these two cities?

5. You have a chemical, "Z." You dissolve it in water and chart your observations. How would you know whether Z is an acid or a base?

Multiple Choice

Circle the letter that best answers the question.

1. Nitrogen has 7 electrons. How many covalent bonds will it form?
   a. 7
   b. 3
   c. 5
   d. 10
   e. 12

2. You discover a new element, Mimionium, which has 400 protons and 401 neutrons. You predict that the atomic form of Mimionium will have _____ electrons and that the negatively charged ion will have _____ electrons.
   a. 801; 401
   b. 401; 399
   c. 400; 401
   d. 401; 800
   e. 400; 399

3. Hydrogen bonding between water molecules occurs between:
   a. the partially positive H of one molecule and a partially negative O of another.
   b. the oxygen atom of one molecule and the hydrogen atom in the same molecule.
   c. the partially negative H of one molecule and a partially positive O of another.
   d. the partially negative H of one molecule and the partially positive H of another.
   e. the partially negative O of one molecule and the partially negative O of another.

4. Which subatomic particle(s) determine the identity of an element?
   a. protons
   b. neutrons
   c. electrons
   d. both protons and neutrons
   e. both neutrons and electrons

5. Which of the following best describes an ion?
   a. an atom in which the numbers of protons and neutrons differ
   b. an atom in which the numbers of protons and electrons differ
   c. an atom in which the numbers of neutrons and electrons differ
   d. an atom in which the number of protons equals the number of electrons
   e. an atom in which the number of protons equals the number of neutrons

6. Which of the following best describes the "rules" for solutions?
   a. Opposites attract.
   b. Generalizations are not useful.
   c. Only solids can be solutes.
   d. Like dissolves like.
   e. Only solids can be solvents.

7. The term used to describe the attraction of an element for electrons is:
   a. solvency.
   b. polarity.
   c. electronegativity.
   d. ionization.
   e. alkalinity.

8. Rank the following terms from least to most inclusive.
   a. neutron, electron, atom, element, matter, solution
   b. matter, electron, atom, neutron, element, solution
   c. electron, neutron, atom, element, solution, matter
   d. atom, electron, neutron, matter, element, solution
   e. element, atom, matter, electron, neutron, solution

9. What is the bond that forms within water molecules called?
   a. ionic
   b. polar covalent
   c. nonpolar covalent
   d. hydrogen
   e. isotopic

10. Molecule A is made up of carbon, hydrogen, and nitrogen atoms. Molecule B contains the same elements in the same proportions, but it has a different three-dimensional shape. What other differences would you expect to find between the two molecules?
    a. Their protons would have different shapes.
    b. Their elements would be different.
    c. Their properties would be different.
    d. Their electrons would be of different types.
    e. They would have different electronegativity values.

11. Free radicals have been blamed for scarring the walls of blood vessels and causing mutational damage to the DNA found in mitochondria. These radicals are formed in response to oxygen's attraction for:
    a. protons.
    b. ions.
    c. electrons.
    d. neutrons.
    e. hydrogen bonds.

12. Had medieval alchemists been successful in changing one element into another, what would they actually have changed about the atoms of the first element? In other words, what truly defines an element?
    a. the number of electron shells
    b. the number of covalent bonds it can form
    c. the number of neutrons
    d. the number of electrons
    e. the number of protons

13. Your new understanding of chemistry has led you to take a job as a research chemist. You discover four new elements with the following properties. Which of these is least likely to form bonds with other elements?
    a. element Jh has four electrons in its outermost shell
    b. element Kd has eight electrons in its outermost shell
    c. element Sb has three electrons in its outermost shell
    d. element Jb has six electrons in its outermost shell
    e. All of these elements are likely to form bonds with other atoms.

14. Why does it take a lot of energy to raise the temperature of water compared to other solvents?
    a. Liquid water is "locked" into a rigid lattice of molecules that doesn't break easily.
    b. Water has a low specific heat.
    c. Energy is needed to break the hydrogen bonds between water molecules.
    d. Water molecules resist the transfer of heat.
    e. Energy is required to break the polar covalent bonds within water molecules.

15. Why do living beings have buffers?
    a. to maintain an acidic pH
    b. to carry out chemical reactions
    c. to maintain a pH close to neutral
    d. to maintain a basic pH
    e. to speed the rate of chemical reactions

16. Hydrangeas are a popular type of garden plant. Certain varieties produce flowers whose color depends on the pH of the soil in which the plant is grown. Soil below a pH of 7 produces blue flowers; soil above 7 produces pink flowers. Which of the following substances would you add to the soil to produce pink flowers?
    a. a base
    b. an acid
    c. a buffer
    d. a solution with a concentration of hydrogen ions that is greater than $1/10^{-7}$
    e. a solution with a concentration of hydrogen ions that is greater than $1/10^{-3}$

17. Which of the following atoms would be considered an ion?
    a. a sodium atom with 11 protons, 12 neutrons, 11 electrons
    b. a chlorine atom with 17 protons, 17 neutrons, 18 electrons
    c. a magnesium atom with 12 protons, 12 neutrons, 12 electrons
    d. a sodium atom with 11 protons, 11 neutrons, 11 electrons
    e. a hydrogen atom with 1 proton and 1 electron

18. Which of the following properties makes water so important to life?
    a. It is a powerful solvent.
    b. Water can absorb and retain heat.
    c. Ice floats.
    d. Water can retain heat.
    e. All of these are properties that make water important to life.

19. You like your coffee black with lots of sugar. What is the solvent?
    a. the sugar
    b. the cup
    c. the cream

d. the coffee

e. the air immediately above the surface of the cup

20. Which of the following would be considered a solute?

   a. salt added to your soup

   b. vinegar used to make salad dressing

   c. milk in your hot chocolate

   d. water used to make lemonade

   e. all of these are solutes

21. Polar and nonpolar covalent bonds both:

   a. involve the exchange of neutrons between their nuclei.

   b. involve the exchange of electrons between their outermost shells.

   c. involve the sharing of electrons between their outermost shells.

   d. involve the loss of electrons from their outermost shells.

   e. none of the above

22. The outermost layer of the atom, where electrons are gained, lost, or shared is the:

   a. proton shell.

   b. electron shell.

   c. valence shell.

   d. neutron shell.

   e. orbital shell.

23. In ionization, atoms:

   a. share the electrons in their outermost shells.

   b. gain electrons.

   c. lose electrons.

   d. gain protons.

   e. lose protons.

24. Suppose that you were given two isotopes of an element to study. How would they differ?

   a. The number of electrons would be different.

   b. The number of protons would be different.

   c. The number of neutrons would be different.

   d. The names would be different.

   e. all of the above

25. When is an atom at its lowest energy state?

   a. when the outermost electron shell is full

   b. when the innermost electron shell is full

   c. when the number of electrons equals the number of neutrons

   d. when the number of protons equals the number of neutrons

   e. when the number of electrons in the innermost shell matches that of the electrons in the outermost shell

WHAT'S IT ALL ABOUT?

Here's a question to help you pull together what you've learned so far using this text.

Question: How does the type of bonding in a molecule affect its properties?

1. **What kind of question is this?**

   This question wants to know how one "thing," in this case bonding, affects another thing, molecular properties.

2. **Collect the evidence.**

   What do we need to know to answer this question?

   a. Types and properties of molecular bonds

   b. Role of bonds in the properties of biological molecules

3. **Pull it all together.**

   Provide an example of each type of bond, and describe how its properties work to hold molecules together. You can use the following topic sentences to get you started.

   • Bonds create stable structures.

   • The solubility of a molecule is a function of its type of bonds.

   • Another property affected by bonding is the three-dimensional shape of a molecule.

CHAPTER 3 LIFE'S COMPONENTS: BIOLOGICAL MOLECULES

Basic Chapter Concepts

- Carbon is a key atom in living organisms because of its unique atomic structure.
- All the important molecules of life—carbohydrates, proteins, lipids, and nucleic acids—use the element carbon as a starting point.
- The function of carbon-based molecules is affected by the addition of specific groups of atoms called functional groups.
- Biomolecules are built from monomeric units—carbohydrates from simple sugars, proteins from amino acids, and nucleic acids from nucleotides.

CHAPTER SUMMARY

3.1 Carbon Is Central to the Living World
- Most biological molecules are built on a carbon framework.
- Carbon is able to form stable covalent bonds with other elements because of its unique electron configuration.
- Carbon bonds readily to oxygen and hydrogen and can make molecules with a variety of shapes.

3.2 Functional Groups
- Specific groups of atoms confer special properties when they are linked to carbon-based molecules.
- Functional groups can affect the bonding capacity of the molecules they are attached to.

3.3 The Molecules of Life: Carbohydrates
- Carbohydrate monomers, called monosaccharides, are composed of carbon, hydrogen, and oxygen.
- These monomers form polymers, called polysaccharides, that can be used as food (such as starch), to store energy (such as glycogen), or to give structure (such as chitin and cellulose).
- Glucose, a six-carbon monosaccharide, is an important energy source for most living things.

3.4 Lipids
- Lipids are a diverse collection of molecules that share the property of being hydrophobic.
- The monomeric unit of all lipids, except the steroids, is the fatty acid—a long chain of carbon and hydrogen atoms ending in a carboxylic acid.
- Some lipids serve as a source and storage form of energy, whereas others, such as steroids and phospholipids, have a structural function.

3.5 Proteins
- Proteins are polymers made from any combination of 20 monomeric units called amino acids (carbon-based molecules that also contain a nitrogen atom).
- Although proteins are linear polymers, they can fold to assume a variety of shapes or conformations. The conformation of a protein is critical to its function.
- Important groups of proteins include enzymes, which speed up chemical reactions, and structural proteins, such as hair and nails.
- Other proteins, called glycoproteins and lipoproteins, are combinations of carbohydrates, proteins, and lipids.

3.6 Nucleic Acids
- Nucleic acids serve a variety of functions, but they are most important for their ability to store and transmit information for constructing protein polymers.
- Nucleic acids, DNA and RNA, are made of nucleotide monomers that contain a phosphate group, a monosaccharide, and a nitrogen-containing base.

WORD ROOTS

carbo- = carbon (e.g., a *carbo*hydrate is composed of carbon, hydrogen, and oxygen)

glyco- = producing sugar (e.g., *glyco*gen is composed of glucose units)

lipo- = fat or other lipids (e.g., *lipo*proteins are compounds composed of lipids and proteins)

mono- = single (e.g., *mono*saccharides are simple sugars)

poly- = many (e.g., a *poly*mer is a substance composed of many small molecules linked together)

KEY TERMS

carbohydrate _____

cellulose _____

chitin _____

cholesterol _____

deoxyribonucleic acid (DNA) _____

fatty acid _____

functional group _____

glycogen _____

glycoprotein _____

hydrocarbon _____

lipid _____

lipoprotein _____

monomer _____

monosaccharide _____

monounsaturated fatty acid _____

nucleotide _____

oil _____

organic chemistry _____

phosphate group _____

phospholipid _____

polymer _____

polypeptide _____

polysaccharide _____

polyunsaturated fatty acid _____

primary structure _____

protein _____

quaternary structure _____

ribonucleic acid (RNA) _____

saturated fatty acid _____

secondary structure _____

simple sugar _____

starch _____

steroid _____

tertiary structure _____

triglyceride _____

wax _____

FLASH CARDS

To use the flash cards, tear the page from the book and cut along the dashed lines. The key term appears on one side of the flash card, and its definition appears on the opposite side.

carbohydrate	**glycoprotein**
cellulose	**hydrocarbon**
chitin	**lipid**
cholesterol	**lipoprotein**
deoxyribonucleic acid (DNA)	**monomer**
fatty acid	**monosaccharide**
functional group	**monounsaturated fatty acid**
glycogen	**nucleotide**

a molecule that combines protein and carbohydrate; glycoproteins play important roles as cell receptors and some types of hormones, among other functions

an organic molecule that always contains carbon, oxygen, and hydrogen and that, in many instances, contains nothing but carbon, oxygen, and hydrogen; carbohydrates usually contain exactly twice as many hydrogen atoms as oxygen atoms

a compound made of hydrogen and carbon; hydrocarbons are nonpolar covalent molecules and therefore are not easily dissolved in water

a complex carbohydrate that is the largest single component of plant cell walls; cellulose is dense and rigid and provides structure for much of the natural world; mammals cannot digest cellulose, so it serves as insoluble dietary fiber that helps move food through the digestive tract

a member of a class of biological molecules whose defining characteristic is their relative insolubility in water; examples include triglycerides, cholesterol, steroids, and phospholipids

a complex carbohydrate that gives shape and strength to the external skeleton of arthropods, including insects, spiders, and crustaceans

a molecule composed of both lipid and protein; lipoproteins transport fat molecules through the bloodstream to all parts of the body

a steroid molecule that forms part of the outer membrane of all animal cells and that acts as a precursor for many other steroids, among them the hormones testosterone and estrogen

a small molecule that can be combined with other similar or identical molecules to make a larger polymer

the primary information-bearing molecule of life, composed of two chains of nucleotides, linked together in the form of a double helix; proteins are put together in accordance with the information encoded in DNA

a building block or monomer of carbohydrates; monosaccharides combine to form complex carbohydrates, or polysaccharides; glucose is an example of a monosaccharide

a molecule, found in many lipids, composed of a hydrocarbon chain bonded to a carboxyl group

a fatty acid with one double bond between the carbon atoms of its hydrocarbon chain

a group of atoms that confers a special property on a carbon-based molecule; functional groups usually are transferred as a unit among carbon-based molecules and often confer an electrical charge or polarity on the molecules they are part of

the building block of nucleic acids, including DNA and RNA, consisting of a phosphate group, a sugar, and a nitrogen-containing base

a complex carbohydrate that serves as the primary form in which carbohydrates are stored in animals

oil	polyunsaturated fatty acid
organic chemistry	primary structure
phosphate group	protein
phospholipid	quaternary structure
polymer	ribonucleic acid (RNA)
polypeptide	saturated fatty acid
polysaccharide	secondary structure

a fatty acid with two or more double bonds between the carbon atoms of its hydrocarbon chain

a dietary lipid that is liquid at room temperature (e.g., olive oil, canola oil)

the sequence of amino acids in a protein; this sequence dictates the final shape of the protein because electrochemical bonding and repulsion forces act on the structure to create the folded-up protein

a branch of chemistry concerned with compounds that have carbon as their central element

a large polymer of amino acids, composed of one or more polypeptide chains; proteins come in many forms, including enzymes, structural proteins, and hormones

a phosphorus atom surrounded by four oxygen atoms

the way in which two or more polypeptide chains come together to form a protein

a charged lipid molecule composed of two fatty acids, glycerol, and a phosphate group; the phospholipid's phosphate group is hydrophilic, whereas its fatty acid chains are hydrophobic; phospholipids are a major constituent of cell membranes

a nucleic acid that is active in the synthesis of proteins and that forms part of the structure of ribosomes; varieties include messenger RNA (mRNA), transfer RNA (tRNA), and ribosomal RNA (rRNA)

a large molecule made up of many similar or identical subunits, called monomers

a fatty acid with no double bonds between the carbon atoms of its hydrocarbon chain

a series of amino acids linked in linear fashion; polypeptide chains fold up to become proteins

the structure proteins assume after having folded up

a polymer of carbohydrates composed of many monosaccharides; examples include starch, glycogen, cellulose, and chitin

simple sugar

tertiary structure

starch

triglyceride

steroid

wax

the large-scale twists and turns in a protein conformation

the smallest and simplest form of carbohydrates, which serve as energy-yielding molecules and as the building blocks or monomers of complex carbohydrates; glucose and fructose are simple sugars

a lipid molecule formed from three fatty acids bonded to glycerol

a complex carbohydrate that serves as the major form of carbohydrate storage in plants; starches—found in such forms as potatoes, rice, carrots, and corn—are important sources of food for animals; in human nutrition, one of the three principal classes of dietary carbohydrates, defined as a complex carbohydrate that is digestible; the other classes of dietary carbohydrates are simple sugars and fibers

a lipid composed of a single fatty acid linked to a long-chain alcohol

a member of the class of lipid molecules that have four carbon rings as a central element in their structure; one steroid differs from another in accordance with the varying side chains that can be attached to these rings; cholesterol, testosterone, and estrogen are steroid molecules

SELF TEST

Once you have finished studying this chapter, close your books, grab a pencil, and spend the next 15 to 20 minutes completing this practice test.

Compare and Contrast

For each of the following paired terms, write a sentence of comparison ("Both") and a sentence of contrast ("However,").

cellulose/starch
polysaccharide/polypeptide
fatty acid/triglyceride
DNA/RNA
monosaccharide/polysaccharide

Short Answer

1. Why do we describe life on Earth as being carbon-based?

2. What do nucleic acids do?

3. Draw a triglyceride, and label the following: glycerol, fatty acids. Where does the condensation reaction occur?

4. If you needed an immediate source of energy, what should you eat, and why?

5. Proteins are notoriously sensitive molecules. Using the protein found in egg white, describe two conditions that might affect the structure of this protein.

6. You have been given the task of creating a new amino acid in the lab, and this novel amino acid is to be used to construct a protein. What two functional groups would you need to use, and how would they be joined to make a protein?

Multiple Choice

Circle the letter that best answers the question.

1. Which of the following macromolecules is/are not technically a polymer?
   a. glucose
   b. glycogen
   c. DNA
   d. steroids
   e. a and d

2. Which of the following protein structures describes the association of two or more proteins?
   a. primary structure
   b. quaternary structure
   c. denaturation
   d. tertiary structure
   e. secondary structure

3. RNA is a polymer of:
   a. amino acids.
   b. glucose.
   c. phospholipids.
   d. steroids.
   e. nucleotides.

4. Which of the following is *not* used as an energy source?
   a. fats
   b. starches
   c. glycogen
   d. oils
   e. nucleic acids

5. When a polymer is formed from two monomers, which of the following is a by-product of the reaction?
   a. ions
   b. proteins
   c. molecule of water
   d. atoms
   e. canned soup

6. The function of phospholipids is to:
   a. form insect shells.
   b. form plant fibers.
   c. form enzymes.
   d. form cell membranes.
   e. make up genetic material.

7. One way to help cure world hunger would be to find an enzyme that would allow humans to digest cellulose. This would work because cellulose is a type of _____, which serves as a source of energy in our bodies.
   a. protein
   b. carbohydrate
   c. lipid
   d. nucleic acid
   e. glue

8. Sickle-cell anemia is a disease that is caused by the change in shape of the hemoglobin molecule. Because hemoglobin is a protein, you know that the mutation that causes this change must occur in the _____ structure of the protein, because it is at this level that three-dimensional shape is determined.
   a. primary
   b. secondary
   c. tertiary
   d. quaternary
   e. quintuple

9. After your yearly physical, you vow to pay more attention to your health. A quick run to the grocery store fills your shopping cart with the following

items. Which one is the best choice for lowering your blood cholesterol levels?
a. a spread comprised of monounsaturated olive oil
b. butter
c. salad dressing containing omega-3 fatty acids
d. an artificial margarine made from trans fatty acids
e. butter and margarine

10. Suppose you have discovered a wonder drug that would cure most of the common human illnesses. Unfortunately, the molecule is hydrophobic and will not dissolve into a solution that can be administered to patients. Which functional group should you add to your molecule to make it more like ethanol, which is water soluble?
a. hydroxyl
b. amino
c. carboxyl
d. phosphate
e. hydrocarbon

11. When you hear the term "organic molecule," the first element that comes to mind is:
a. hydrogen.
b. titanium.
c. carbon.
d. oxygen.
e. nitrogen.

12. A –COOH group is a(n):
a. hydroxyl.
b. amino.
c. phosphate.
d. carboxyl.
e. alcohol.

13. When a hydroxyl group is added to a hydrocarbon group, it becomes a(n):
a. solid.
b. fat.
c. alcohol.
d. gas.
e. semisolid.

14. French fries are a favorite American snack food. They contain which of the following two bioorganic molecules?
a. glucose and sucrose
b. salt and ketchup
c. carbohydrates and lipids
d. starch and potato skins
e. carbohydrates and proteins

15. Lipids are different from other biologic molecules because they:
a. are not made of specific monomeric units.
b. are mostly insoluble in water.
c. are made of carbon, hydrogen, and oxygen.

d. a and b
e. a and c

16. Which of the following is *not* a complex carbohydrate?
a. glucose
b. cellulose
c. sucrose
d. maltose
e. glycogen

17. The critical energy-rich molecule that is carried in your blood is _____; when the level of this molecule increases, your liver converts this molecule into _____.
a. glycogen; glucose
b. glucose; glycogen
c. sucrose; sucrase
d. lipids; fats
e. fats; lipids

18. Carbon has four electrons in its valence shell. What types of bonds would you expect it to form?
a. hydrogen
b. covalent and ionic
c. covalent
d. ionic
e. hydrophobic

19. Estrogen, testosterone, and cholesterol are examples of:
a. polysaccharides.
b. nucleic acids.
c. lipids.
d. food additives.
e. fats.

20. Antigen and antibodies have important roles in the defense against invaders. They are examples of:
a. hormones.
b. proteins with a complex three-dimensional shape.
c. enzymes.
d. lipoproteins.
e. proteins that provide mechanical support.

21. Alcohol can be used as an antibacterial agent because it:
a. contains hydroxyl groups.
b. denatures bacterial proteins.
c. denatures bacterial DNA.
d. dries up bacteria.
e. is hydrophobic.

22. Daffodils are lovely spring flowers that are well known for their brilliant yellow color. The flower color is determined by instructions in its:
a. proteins.
b. nucleic acids.
c. carbohydrates.

d. lipids.

e. none of the above

23. Students often gain weight during their first semester of college. You could explain this to your family by commenting: "I'm storing up some _____ for winter."

a. polysaccharides

b. triglycerides

c. nucleotides

d. polypeptides

e. steroids

24. You are not sure whether a molecule is a protein or a nucleic acid. One thing that they both have in *common* is that they:

a. are both hydrophobic.

b. are both made of amino acids.

c. are both polymers.

d. are both excellent energy sources.

e. each consist of four basic kinds of subunits.

25. In the game of BioJeopardy, you have chosen the category of biological molecules. The answer is: "This carries oxygen in the blood." What is the question?

a. What is hemoglobin?

b. What is a steroid?

c. What is an antibody?

d. What is an antigen?

e. What is glucose?

WHAT'S IT ALL ABOUT?

Here's a question to help you pull together what you've learned so far using this text.

Question: Most biologically important molecules are polymers built from monomeric units. Why is this advantageous to a living organism? Support your answer with examples.

1. **What kind of question is this?**

   This is a variation on the "defend a statement" question. You need to provide evidence to support the claim that making large molecules out of "building blocks" provides some advantage to the organism.

2. **Collect the evidence.**

   "Advantages" are general properties of molecules that serve a purpose. To document that these properties are advantages, you need to find specific examples of how a polymer can be more functional than its monomeric unit.

3. **Pull it all together.**

   Each advantage that you identify will be the topic of a paragraph. The body of each paragraph will provide specific examples of how building polymers from monomers produces the advantage you claim. You can use the following topic sentences to get you started.

   • Creating polymers from monomer offers two advantages to the organism: ease of molecular construction and increased diversity of molecules.

   • In addition to easy construction, the cell can make many different molecules from the same collection of monomers just by changing the order of the monomers.

CHAPTER 4 LIFE'S HOME: THE CELL

Basic Chapter Concepts

- All cells are either prokaryotic (bacteria and archaea) or eukaryotic (protists, fungi, plants, and animals).
- The two types of eukaryotic cells—animal cells and plant cells—possess a nucleus and other intracellular organelles that allow for compartmentalization of cellular activities.
- Plant cells differ from animal cells because they have (1) cell walls, (2) a central vacuole for water storage, and (3) plastids such as chloroplasts.

CHAPTER SUMMARY

4.1 Cells Are the Fundamental Units of Life
- All living organisms are composed of cells that share a fundamental unity of structure.
- All cells have at least one membrane that separates the contents of the cell from its surroundings.
- Cells contain specialized molecules that permit the reactions of life to occur.
- All cells can make copies of themselves.

4.2 All Cells Are Either Prokaryotic or Eukaryotic
- The prototypical prokaryotic cell is a bacterium—a single-celled organism that lacks intracellular organelles. Prokaryotic cells are smaller and functionally simpler than eukaryotic cells.
- Compartmentalization of specialized functions within membrane-bound organelles of eukaryotic cells facilitates formation of tissues and organs.
- Both plants and animals are made up of eukaryotic cells. Plant cells differ from animal cells somewhat in their complement of organelles, but both types of eukaryotic cells share many structures and processes.

4.3 The Eukaryotic Cell
- The eukaryotic cell can be subdivided into five functional units: the nucleus; the cytosol, which contains the organelles; the cytoskeleton; and the plasma membrane.

4.4 A Tour of the Animal Cell: Along the Protein Production Path
- The nucleus contains the genetic material (DNA) that controls the structure and reproduction of the cell through the synthesis of proteins. The nucleus also contains the nucleolus, which makes the RNA molecules that will become the ribosomes.
- Proteins move around within the cell, and from the inside to the outside of the cell, through the action of an endomembrane system consisting of the rough endoplasmic reticulum (RER) and the Golgi complex.
- Proteins made in the RER are packaged into vesicles that can move through the cytoplasm to an appropriate destination, or that can fuse with the plasma membrane and release their contents to the outside.

4.5 Outside the Protein Production Path: Other Cell Structures
- Organelles—such as mitochondria, endoplasmic reticulum, and lysosomes—are membrane-delimited structures that allow for cellular activities to be isolated from each other.
- Mitochondria produce energy for the cell, whereas the acidic lysosomes break down and recycle cellular molecules; membranes prevent the work of one set of reactions from interfering with another set of reactions.
- Organelles are suspended in a protein-rich, aqueous environment called the cytosol.
- Cells are not just sacks of structures; cells have architecture provided by the internal scaffolding of the cytoskeleton. The cytoskeleton gives animal cells a flexible shape.

4.6 The Cytoskeleton: Internal Scaffolding
- Three types of protein structures make up the cytoskeleton. (1) Microfilaments are the smallest and are made up of actin. (2) Intermediate filaments are larger and provide support and structure within the cell. (3) The largest, microtubules, also provide support and movement both inside and outside the cell.

4.7   The Plant Cell

- The primary feature that distinguishes a plant cell from an animal cell is the cell wall, a thick, rigid structure composed mostly of cellulose.
- The cell wall gives plants their distinctive shape and also limits water movement into and out of the cell.
- The central vacuole dominates the cytoplasm of the plant cell. This vacuole is a primary site for metabolism because it stores nutrients, processes waste products, stores pigment molecules, and controls the internal pH of the cell.
- Plant cells contain organelles called plastids. Plastids also store pigments and process nutrients. The best-known plastids are the chloroplasts, the site of photosynthesis in green plants and algae.

4.8   Cell-to-Cell Communication

- Multicellularity is no advantage if each cell still functions as a separate entity.
- Cells within tissues and organs communicate with each other through channels.
- Plant-cell channels are called plasmodesmata and must penetrate the cell wall.
- Animal-cell channels are called gap junctions.
- Plasmodesmata and gap junctions are specialized protein structures that control the flow of molecules and electrical signals between cells.

WORD ROOTS

cyto- = cell (e.g., the *cyto*skeleton is the support network within the cell)

lyso- = dissolve (e.g., *lyso*some breaks down cellular debris)

micro- = small (e.g., *micro*scopes allow us to view structures too small to be seen by the naked eye)

nano- = one billionth (e.g., a *nano*meter is one billionth [1/1,000,000,000 or 10^{-9}] of a meter)

KEY TERMS

cell wall _____

_____

central vacuole _____

_____

chloroplast _____

_____

cilia _____

_____

cytoplasm _____

_____

cytoskeleton _____

_____

cytosol _____

_____

endomembrane system _____

_____

eukaryotic cell _____

_____

flagella _____

_____

gap junction _____

_____

Golgi complex _____

_____

intermediate filament _____

_____

lysosome _____

_____

microfilament _____

_____

micrograph _____

_____

micrometer _____

_____

microtubule _____

_____

mitochondria _____

_____

nanometer _____

_____

nuclear envelope _____

_____

nucleolus _____

_____

nucleus _____

_____

organelle _____

_____

plasma membrane _____

_____

plasmodesmata _____

_____

prokaryotic cell _____

_____

ribosome _____

_____

rough endoplasmic reticulum _____

_____

smooth endoplasmic reticulum _____

_____

transport vesicle _____

_____

SELF TEST

Once you have finished studying this chapter, close your books, grab a pencil, and spend the next 15 to 20 minutes completing this practice test.

Compare and Contrast

For each of the following paired terms, use a sentence of comparison ("Both") and a sentence of contrast ("However,").

cilia/flagella
nucleus/nucleolus
cytosol/cytoplasm
smooth ER/rough ER
prokaryote/eukaryote

Short Answer

1. Plants have plasmodesmata, and animals have gap junctions for communication. Why is this ability important for multicellular organisms?

2. Why should we not be surprised that plant cells have central vacuoles and walls and that animal cells don't?

3. If the cell is a factory, churning out materials, what factory part or function is analogous to the mitochondria?

4. What would happen to the proteins synthesized on the rough ER if the Golgi complex were not present?

5. Draw the sequence of structures required for export protein synthesis within the cell.

6. Explain the significance of the nucleus in a eukaryotic cell.

7. List and draw the major organelles of a prokaryotic cell.

8. Explain the significance of the mitochondria.

Multiple Choice

Circle the letter that best answers the question.

1. The smallest of the cytoskeleton elements is the:
   a. microtubule.
   b. intermediate filament.
   c. micromini.
   d. microfilament.
   e. flagella.

2. The largest structure visible within an animal cell is usually the:
   a. mitochondria.
   b. Golgi apparatus.
   c. ribosomes.
   d. nucleus.
   e. cytosol.

3. Which of the following is the building site for proteins?
   a. nucleolus
   b. smooth endoplasmic reticulum
   c. ribosome
   d. DNA
   e. mitochondria

4. Which of the following is a site for lipid synthesis?
   a. lysosomes
   b. rough endoplasmic reticulum
   c. smooth endoplasmic reticulum
   d. nucleus
   e. plasma membrane

5. Which organelle is the recycling center of the cell?
   a. mitochondria
   b. the nucleus
   c. smooth ER
   d. the lysosome
   e. the Golgi body

6. What types of eukaryotic cells are flagellated?
   a. motile bacteria
   b. all eukaryotic cells
   c. sperm
   d. no eukaryotic cells
   e. jellyfish cells

7. Which of the following structures permit(s) communication between eukaryotic cells?
   a. gap junctions
   b. plasmodesmata
   c. central vacuole
   d. microfilaments
   e. a and b

Use the following description to answer questions 8–12.

After collecting samples of the material found growing on last month's leftovers at the back of your fridge, you realize that you have discovered a new type of organism. You find that the cells have distinct subunits within. Further, several of these subunits, which you have nicknamed Groucho, Chico, and Harpo, seem to be involved in the production and transfer of certain molecules. Molecule Dumont, as you call it, appears after a messenger molecule is sent from Groucho to Chico. Dumont then travels to Harpo, where it is covered in membrane and (eventually) released from the cell.

8. What type of cells do you think these are?
   a. prokaryote
   b. eukaryote
   c. bacteria
   d. virus
   e. There is insufficient evidence to judge; any of the above organisms use this system.

9. Dumont is most likely what type of molecule?
   a. protein
   b. lipid
   c. carbohydrate
   d. DNA
   e. water

10. Which of the following is the likely nucleus of these cells?
    a. Harpo
    b. Groucho
    c. Chico
    d. Dumont
    e. Zeppo

11. Which of the following has the same function as the Golgi complex?
    a. Harpo
    b. Groucho
    c. Chico
    d. Dumont
    e. Zeppo

12. Which of the following is *not* a eukaryote?
    a. fungus
    b. dog
    c. tree
    d. bacterium
    e. bird

13. Which of the following is characteristic of all prokaryotes?
    a. no nucleus
    b. multicellular
    c. a true nucleus
    d. a Golgi complex
    e. membrane-bound organelles

14. Which of the following is common to both plant and animal cells?
    a. cell wall
    b. organelles
    c. central vacuole
    d. mitochondria
    e. chlorophyll

15. What is the structural difference between rough ER and smooth ER?
    a. Rough ER has ribosomes, but smooth ER does not.
    b. Smooth ER has DNA, but rough ER does not.
    c. Smooth ER has a lipid membrane, but rough ER does not.
    d. Rough ER is found within the nucleus; smooth ER is in the cytoplasm.
    e. Smooth ER is found within the nucleus; rough ER is in the cytoplasm.

16. Which organelle provides the energy you use to read this sentence?
    a. nucleus
    b. central vacuole
    c. mitochondria
    d. microtubule
    e. smooth ER

17. Gap junctions and plasmodesmata have several features in common. What are they?
    a. They both allow material to move from one cell to another.
    b. They are both made up of proteins.
    c. They are both found in plants.
    d. They are both found in animals.
    e. They are both used to anchor cells to underlying tissues.

18. DNA is located in the _____, and its message reaches the ribosome by _____.
    a. cytoplasm; vesicles
    b. nucleus; vesicles
    c. cell membrane; nuclear pores
    d. nucleus; nuclear pores
    e. cytoplasm; microfilaments

19. Which of the following lists the structures involved in protein synthesis and release, in the correct order?
    a. ribosome, mRNA, smooth ER, nucleus
    b. smooth ER, nucleus, ribosome, mRNA
    c. nucleus, mRNA, ribosome, smooth ER
    d. mRNA, nucleus, ribosome, smooth ER
    e. mRNA, ribosome, nucleus, smooth ER

20. Which of the following lists the cytoskeletal elements from smallest to largest?
    a. microfilament, intermediate filament, microtubule
    b. microtubule, microfilament, intermediate filament
    c. microfilament, microtubule, intermediate filament
    d. intermediate filament, microfilament, microtubule
    e. intermediate filament, microtubule, microfilament

21. Cilia and flagella are made up of:
    a. microfilaments.
    b. microtubules.
    c. intermediate filaments.
    d. hemoglobin.
    e. centrioles.

22. Chloroplasts are found in:
    a. plants.
    b. animals.
    c. prions.
    d. yeast.
    e. viruses.

23. Ribosomal RNA is synthesized in the:
    a. cell wall.
    b. smooth ER.
    c. nucleolus.
    d. ribosome.
    e. cytoskeleton.

24. Ribosomes "dock" with rough ER:
- **a.** to produce proteins that will remain in the cell's cytoplasm.
- **b.** to produce proteins that will used within the cell's nucleus.
- **c.** to produce proteins that will be exported from the cell.
- **d.** to produce carbohydrates for cell metabolism.
- **e.** to produce lipids that will maintain the lipid bilayer.

25. Which of the following eukaryotic organelles is specific to plant cells?
- **a.** I
- **b.** II
- **c.** III
- **d.** IV
- **e.** V

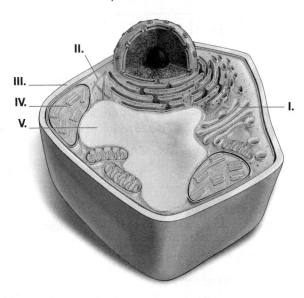

From the list below, fill the correct term in the blank spaces provided on the figure of the animal cell below.

mitochondria	nuclear envelope	DNA
nucleus	cytoskeleton	plasma membrane
rough endoplasmic reticulum	free ribosomes	nuclear pores
lysosomes	transport vesicle	nucleolus
Golgi complex	cytosol	smooth endoplasmic reticulum

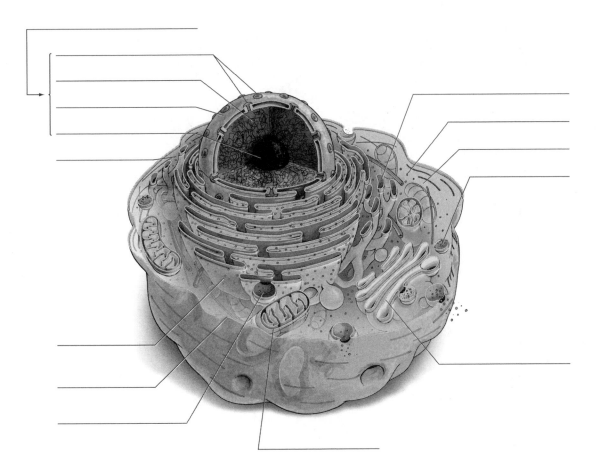

WHAT'S IT ALL ABOUT?

Here's a question to help you pull together what you've learned so far using this text.

Question: In this chapter, we've learned that all living things are cells or are made of cells. Because we know that there are four classes of biomolecules (proteins, lipids, carbohydrates, and nucleic acids), then it follows that cells and the structures within the cells must be made from these biomolecules. How do these molecules shape the structure and the function of cellular components?

1. **What kind of question is this?**

   This question wants to know how one "thing," in this case a type of molecule, impacts two other things, higher-order cellular structure and the functioning of those structures.

2. **Collect the evidence.**

   What do we need to know to answer this question?
   a. Molecular composition of cellular structures
   b. Function of cellular structures

3. **Pull it all together.**

   For each class of biomolecule, provide an example of a type of molecule that makes up a particular cellular structure, and describe why this type of molecule is important to the "job" of that structure. You can use these first two sentences to get you started.

   All parts of the cell are made from the four basic types of molecules: lipids, proteins, nucleic acids, and carbohydrates. Each class of biomolecules contributes to the specialized functions of the cell's component parts.

CHAPTER 5 LIFE'S BORDER: THE PLASMA MEMBRANE

Basic Chapter Concepts

- The plasma membrane is the outermost boundary of all cells. It has multiple roles as protector, gatekeeper, and passageway for cellular messages.
- The functionality of the plasma membrane relies upon its structure.
- Water moves freely across the membrane, down a concentration gradient, in a process called osmosis. Some small molecules also freely diffuse through the cell membrane.
- Transport of most molecules across the membrane requires specialized membrane structures. The process can be either passive, down a concentration gradient; or active, against a concentration gradient.
- Very large molecules must be packaged to move into or out of the cell. These processes are called endocytosis and exocytosis, respectively.

CHAPTER SUMMARY

5.1 The Nature of the Plasma Membrane

- The plasma membrane functions first as a barrier—it keeps biological molecules concentrated inside the cell and keeps harmful substances outside.
- Membranes are made of a bilayer of phospholipids. The phospholipids are oriented such that their hydrophobic heads face the aqueous environment (on the inside and outside of the cell), whereas their hydrophobic tails face each other.
- Other lipid molecules, such as cholesterol, help control the fluidity of the membrane.
- Proteins within the lipid bilayer allow selective movement of molecules into and out of the cell. Proteins may also interact with cytosolic molecules and help determine the shape of the membrane.
- Various molecules embedded in the membrane receive information from signal molecules outside the cell.

5.2 Diffusion, Gradients, and Osmosis

- Diffusion occurs as molecules randomly move from an area of high concentration to one of low concentration. When the molecules moving are water, we call the process osmosis.
- The cell membrane is a semipermeable barrier; water can move across it freely despite its polarity, but most solutes cannot. This selective movement of solutes creates a concentration gradient between the inside and outside of the cell.
- When the movement of solutes is limited, water moves through the cell membrane (osmosis) to balance the solute concentration inside and outside the cell.

5.3 Moving Smaller Substances In and Out

- Passive transport requires no input of energy; molecules simply move down their concentration gradient. Large molecules that cannot pass through the membrane are facilitated in diffusion by protein channels within the membrane.
- Active transport requires energy because molecules are being moved against their concentration gradient—these molecules are being concentrated either outside or inside the cell.

5.4 Getting the Big Stuff In and Out

- Exocytosis is the outward movement of large molecules. Exported molecules, such as proteins, are made within the cell and packaged into lipid vesicles that diffuse through the cytoplasm and fuse with the membrane. This fusion dumps the contents outside the cell.
- Endocytosis moves molecules into the cell. Endocytosis requires the plasma membrane to invaginate, creating a "bubble" that contains extracellular material.
- Small endocytotic vesicles collect water and solutes; this process is called pinocytosis, or "cell drinking." Larger vesicles form when extracellular receptors bind their ligands, then collect in one spot on the

membrane and form a pit. This pit invaginates and internalizes both the receptor and ligand in a process called receptor-mediated endocytosis.

- Some cells can engulf very large items, such as whole bacteria. The cellular membrane pushes out pseudopodia that surround and capture the item for the cell in a very large (1–2 mm) vesicle. This process is called phagocytosis, or "cell eating."

WORD ROOTS

endo- = internal or within (e.g., *endo*cytosis is the uptake of substances into cells)

exo- = external or from outside (e.g., *exo*cytosis is the movement of substances out of cells)

hyper- = over or above (e.g., a *hyper*tonic solution is a solution whose concentration of solutes is higher than that found in cells)

iso- = equal (e.g., an *iso*tonic solution is one in which the concentration of solutes is equal both inside and outside the cell)

phago- = to eat (e.g., *phago*cytosis is the "eating" of bacteria, food particles, and other debris by specialized cells)

pino- = to drink (e.g., *pino*cytosis is the "drinking" or taking in of liquids by a cell)

KEY TERMS

active transport _____

concentration gradient _____

diffusion _____

endocytosis _____

exocytosis _____

facilitated diffusion _____

fluid-mosaic model _____

glycocalyx _____

hypertonic solution _____

hypotonic solution _____

integral protein _____

isotonic solution _____

osmosis _____

passive transport _____

peripheral protein _____

phagocytosis _____

phospholipid bilayer _____

pinocytosis _____

plasma membrane _____

receptor protein _____

simple diffusion _____

transport protein _____

FLASH CARDS

To use the flash cards, tear the page from the book and cut along the dashed lines. The key term appears on one side of the flash card, and its definition appears on the opposite side.

active transport	fluid-mosaic model
concentration gradient	glycocalyx
diffusion	hypertonic solution
endocytosis	hypotonic solution
exocytosis	integral protein
facilitated diffusion	isotonic solution

a conceptualization of the cell's plasma membrane as a fluid, phospholipid bilayer that has within it a mosaic of both stationary and mobile proteins

---

transport of materials across the plasma membrane in which energy is expended; through active transport, solutes can be moved against their concentration and electrical gradients; the sodium-potassium pump is an example of active transport

---

an outer layer of the plasma membrane, composed of short carbohydrate chains that attach to membrane proteins and phospholipid molecules; such chains serve as the actual binding sites on many membrane proteins, act to lubricate the cell, and can form an adhesion layer that allows one cell to stick to another

---

a gradient within a given medium defined by the difference between the highest and lowest concentration of a solute; the solute will have a natural tendency to move from the areas of higher concentration to lower, thus diffusing

---

a solution that has a high concentration of solutes relative to an adjacent solution

---

the movement of molecules or ions from areas of their higher concentration to areas of their lower concentration; over time, the random movement of molecules will result in the even distribution of the material

---

a solution that has a low concentration of solutes relative to an adjacent solution

---

the process by which cells bring relatively large materials into themselves through use of transport vesicles

---

a protein of the plasma membrane that is attached to the membrane's hydrophobic interior

---

the means by which relatively large volumes of material are moved from the inside of a cell to the outside; in exocytosis, a transport vesicle fuses with a cell's plasma membrane, after which the contents of the vesicle are ejected outside the cell

---

a solution that has the same concentration of solutes as an adjacent solution

---

a passage of materials through the cell's plasma membrane that is aided by a transport protein

osmosis

pinocytosis

passive transport

plasma membrane

peripheral protein

receptor protein

phagocytosis

simple diffusion

phospholipid bilayer

transport protein

the movement of relatively large materials into a cell by means of the creation of transport vesicles that are produced through an invagination of the plasma membrane; one of two primary forms of endocytosis, the other being phagocytosis

the net movement of water across a semipermeable membrane from an area of lower solute concentration to an area of higher solute concentration

a membrane forming the outer boundary of many cells; composed of a phospholipid bilayer interspersed with proteins and cholesterol molecules and coated, on its exterior face, with short carbohydrate chains associated with proteins and lipids

transport of materials across the cell's plasma membrane that involves no expenditure of energy; simple and facilitated diffusion are examples of passive transport

a protein that protrudes from the plasma membrane of a cell that is active in the transmission of signals between cells; receptor proteins, often called simply *receptors*, function by binding with signaling molecules such as hormones

a protein of the plasma membrane that lies on the inside or outside of the membrane but is not attached to the membrane's hydrophobic interior

diffusion through the plasma membrane that requires only concentration gradients, as opposed to concentration gradients and special protein channels; water, oxygen, carbon dioxide, and steroid hormones can all cross the plasma membrane through simple diffusion

the movement of large materials into a cell by means of wrapping extensions of the plasma membrane around the materials and fusing the extensions together; one of two primary forms of endocytosis, the other being pinocytosis

a protein that forms a hydrophilic channel through the hydrophobic interior of the cell's plasma membrane, allowing hydrophilic materials to pass through the membrane

one of the chief components of the plasma membrane, composed of two layers of phospholipids, arranged with their fatty acid chains pointing toward each other

SELF TEST

Once you have finished studying this chapter, close your books, grab a pencil, and spend the next 15 to 20 minutes completing this practice test.

Compare and Contrast

For each of the following paired terms, write a sentence of comparison ("Both") and a sentence of contrast ("However,").

phospholipids/cholesterol
passive diffusion/facilitated diffusion
phagocytosis/pinocytosis
integral/peripheral proteins
hypertonic/hypotonic

Short Answer

1. What specific function of the cell membrane is defective in a cystic fibrosis patient?

2. What functions do proteins carry out on the external surface of the cytoplasmic membrane?

3. If shipwrecked sailors die after drinking lots of highly concentrated seawater, what do you predict would happen to the cells of a person after drinking copious quantities of distilled water? (Distilled water has no dissolved solutes of any kind.)

4. Which of the following substances would you expect to cross the membrane "solo"? For each, explain your answer.

Substance	Cross Solo?	Reason
Water		
Oxygen		
A large protein		
A small molecule with a large charge		
A small molecule with a large charge		

Multiple Choice

Match the following terms with their description. Each choice may be used once, more than once, or not at all.

   a. integral protein
   b. peripheral protein
   c. cholesterol
   d. phospholipid
   e. carbohydrate ("sugar")

1. _____ Found as a bilayer

2. _____ Maintains membrane fluidity and serves as a patching compound

3. _____ Has ends that stick out of both sides of membrane

4. _____ Is both "water loving" and "water fearing"

5. _____ Serves as a binding site for proteins on the cell surface

6. Which of the following best describes the process of facilitated diffusion?
   a. movement of water from areas of high solute concentration to areas of lower solute concentration
   b. expenditure of energy to move solutes across a membrane against their concentration gradients
   c. expenditure of energy to move solutes across a membrane down their concentration gradients
   d. movement of solutes across a membrane down their concentration gradient with the involvement of membrane proteins
   e. movement of solutes across a membrane down their concentration gradient in the absence of membrane proteins

7. Which of the following statements concerning the plasma membrane is *false*?
   a. It is composed of a phospholipid bilayer.
   b. It has integral and peripheral proteins.
   c. It forms the border between the cell and the external environment.
   d. Cholesterol helps add a rigid framework, or scaffold, to the membrane.
   e. Composition can vary in different organisms.

8. You have mutated an organism (Larklula) to produce chocolate candy within its cells. But Larklula is useless unless it can release the candy from its cells so that you can harvest it. Which of the following processes will Larklula need to use to accomplish this?
   a. osmosis
   b. phagocytosis
   c. exocytosis
   d. pinocytosis
   e. halitosis

9. Which of these is *not* a function of membrane proteins?
   a. structure
   b. recognition
   c. reproduction
   d. communication
   e. transport

10. Which of the following features explains why plant cells don't explode as a result of osmosis?
    a. Chloroplasts produce sugar, which prevents water from being drawn into the cell.
    b. Plant cells are all dead at maturity.
    c. Plant cells have a large reservoir, called a central vacuole, where they keep their water.
    d. Plant cells have a rigid cell wall.
    e. Plant cells are more likely than animal cells to explode because of osmosis.

11. To build tissues, it is necessary to bind cells together by attaching membrane proteins and phospholipids. Short chains of what kind of molecule form this glycocalyx?
    a. carbohydrate
    b. protein
    c. phospholipid
    d. nucleic acid
    e. fats

12. Which of the following statements is *not* a critical function of cell membranes?
    a. holding necessary substances within the cell
    b. preventing invasion of harmful material
    c. releasing energy for building new structures
    d. allowing passage of material into and out of the cell
    e. water transport

13. What is a concentration gradient?
    a. a difference in the density of substances from one side of a structure to the other
    b. a way of measuring depth perception
    c. a type of protein used to bring material into the cell
    d. the amount of energy required to run the sodium-potassium pump in active transport
    e. the amount of molecules within a cell

14. You are given two substances to study, one large and one small. Given what you have learned about the movement of material across membranes, which of the following methods would you predict could be used by cells to transport the larger substance?
    a. diffusion
    b. osmosis
    c. active transport
    d. exocytosis
    e. facilitated transport

15. You observe a cell as it takes up certain molecules. You notice that transport depends on the interaction of substance X with a membrane protein. You conclude that the mode of cell uptake being demonstrated is:
    a. receptor-mediated endocytosis.
    b. facilitated diffusion.
    c. phagocytosis.
    d. pinocytosis.
    e. exocytosis.

16. Let's say you find yourself in an alternate universe, where lipid is the universal solvent and comprises approximately 70 percent of your body. What properties would the molecules that make up your cellular membranes have?
    a. They are likely to be lipid soluble.
    b. They are likely to be water soluble.
    c. They would serve as a semipermeable membrane, allowing only certain materials to freely cross the membrane.
    d. They are likely to be hydrophilic.
    e. They would probably have a higher concentration of carbohydrates.

17. You decide to try your hand at canning pickles. You immerse freshly picked cucumbers in a solution that has a solute concentration twice that found in the cucumber cells. You allow your preparation to cure for several months in a sealed jar. When you open the jar later, you find that the fluid surrounding the "pickle" is more dilute than when you started. This change in concentration is due to:
    a. water leaving the cucumber along its concentration gradient.
    b. water traveling against its concentration gradient.
    c. solute leaving the cucumber.
    d. solute entering the cucumber.
    e. not boiling the cucumbers.

18. A nursing baby is able to obtain necessary proteins from its mother's milk. These molecules are able to enter the cells lining the baby's digestive tract through:
    a. osmosis.
    b. passive transport.
    c. active transport.
    d. endocytosis.
    e. exocytosis.

19. The fluid mosaic model was ultimately the best depiction of the cell membrane because:
    a. it showed the membrane as a sandwich, with lipid on the outside and protein in the middle.
    b. it showed the membrane as having a hydrophilic layer on the outside and a hydrophobic layer on the inside.
    c. it proposed that a layer of proteins coated the outside of the membrane.
    d. it showed that the phospholipids and proteins within the layers were able to move.
    e. it showed the membrane as being made of mostly carbohydrates, with few lipids and proteins.

20. If phospholipids are dropped into water:
  a. they will immediately form a cell membrane.
  b. they will disperse as tiny droplets.
  c. a structure will form with the hydrophobic tails pointing outward and the hydrophilic heads pointing inward.
  d. they will disappear.
  e. a structure will form with the hydrophilic tails pointing outward and the hydrophobic heads pointing inward.

21. When a white blood cell munches up a bacterium, this is an example of:
  a. endocytosis.
  b. phagocytosis.
  c. fast food.
  d. exocytosis.
  e. osmosis.

22. If you are shipwrecked, it is *not* a good idea to drink seawater because:
  a. it will just make you thirstier, because it is a hypertonic solution.
  b. it is contaminated with bacteria.
  c. it is hypotonic to your body fluids.
  d. it should be boiled first.
  e. it will cause your cells to swell because it is a hypotonic solution.

23. Although we are cautioned against eating foods that are high in cholesterol, we need a small amount of cholesterol because:
  a. it is part of the cell membrane.
  b. it helps keep cell membranes fluid.

  c. it helps keep cell membranes solid.
  d. without it, we would develop atherosclerosis.
  e. if we do not consume enough cholesterol, our cells cannot form membranes.

24. In cell transport, the protein clathrin is involved in:
  a. active transport.
  b. phagocytosis.
  c. endocytosis.
  d. capturing bacteria.
  e. diffusion.

25. You just bought a new plant fertilizer called Plant Zap, and you immediately apply it to your African violets. After several hours, you notice that the plant is drooping over! Which of the following best hypothesizes what could have happened?
  a. The fertilizer was too concentrated and caused the plant cells to shrink.
  b. The fertilizer was too dilute and caused the plant cells to burst.
  c. The fertilizer was hypertonic to the plant cells.
  d. You didn't read the directions telling you to dilute the fertilizer before using.
  e. The fertilizer was isotonic to the plant cells.

WHAT'S IT ALL ABOUT?

Here's a question to help you pull together what you've learned so far using this text.

Question: Back in Chapter 1, we considered how to define living things. We considered such features as assimilating and using energy, responding to the environment, reproducing, and evolving from other living things. Whereas all of these definitions are helpful, we suggest that the simplest way to distinguish living and nonliving things is by the presence of plasma membranes (or one plasma membrane). Is this a reasonable statement and why?

1. **What kind of question is this?**
   This question asks you to defend the statement that only living organisms have plasma membranes.

2. **Collect the evidence.**
   What do we need to know to answer this question?

   We know that rocks, water, and soil lack plasma membranes. The object is to describe how the plasma membrane makes these functions (such as responding to the environment) possible.

3. **Pull it all together.**
   Select several features of living organisms, and provide examples of how the plasma membrane helps the organism accomplish these tasks on a cellular level. You can use the following sentence to get you started: Because plasma membranes allow living organisms to do the things they do, we can argue that the simplest definition of a living thing is something that has plasma membranes.

CHAPTER 6 LIFE'S MAINSPRING: AN INTRODUCTION TO ENERGY

Basic Chapter Concepts

- Living organisms capture energy from the sun and use it to carry out complex activities, with some energy lost as heat.
- Energy is the capacity to move against an opposing force; energy can be transformed but is not lost or gained (first law of thermodynamics).
- Energy appears in biological systems in the form of the molecule ATP; the energy stored in its chemical bonds can be transformed to cause movement, to transport molecules, or to make other chemical bonds.
- Enzymes are proteins that speed up and regulate energy transformations in living organisms.

CHAPTER SUMMARY

6.1 Energy Is Central to Life
- The ultimate source of energy for almost all living organisms is the sun.

6.2 The Nature of Energy
- We define energy by its effect. Energy cannot be isolated, because it is an intrinsic property of a system. We distinguish the potential of a system to create change as potential energy, whereas energy seen as motion is kinetic energy.
- The first law of thermodynamics allows energy to be transformed. Solar energy can be converted into chemical energy, and chemical energy can be converted into motion. However, energy cannot be created or destroyed.
- The second law of thermodynamics tells us that giving up energy means moving from an ordered state to a less-ordered state; entropy is a measure of the degree of order in a system.

6.3 How Is Energy Used by Living Things?
- Living systems use energy to do three kinds of work: mechanical, transport, and synthetic.
- Each kind of work creates a local increase in the orderliness of the organism; such processes require the input of energy and are called endergonic reactions.
- Exergonic reactions release energy and create a decrease in the orderliness of the organism; this released energy powers the endergonic reactions as a coupled system.

6.4 The Energy Dispenser: ATP
- ATP (adenosine triphosphate) is a relatively unstable molecule that releases large amounts of energy when it loses a phosphate group, which can be harnessed to run an endergonic reaction.
- ATP doesn't store energy for long periods of time; instead, it transfers it from one molecule to another during phosphate transfer.

6.5 Efficient Energy Use in Living Things: Enzymes
- To carry out the complex activities of life, organisms must have a way to control the capture and release of energy.
- Specialized molecules called enzymes speed up virtually all biological reactions by lowering the activation energy barrier.
- Enzymes are not consumed by biological reactions; they are catalysts that are restored at the end of the reaction and can be reused.

6.6   Lowering the Activation Barrier through Enzymes
- Enzymes are catalysts. Catalysts increase the rate at which a reaction occurs without being used up themselves.
- The amount of energy required to start a reaction is called the activation energy. Enzymes decrease the amount of activation energy required, thereby lowering the amount of energy that is needed to start a reaction.
- Enzymatic reactions take place at the active site on the enzyme, where substrate transformation occurs.

6.7   Regulating Enzymatic Activity
- Enzyme activity can be controlled by the presence of regulator molecules to slow down or speed up reactions, in a process known as allosteric regulation.
- Allosteric regulators bind to the enzyme, but not at the active site. The subsequent change in molecular shape changes the ability of the substrate to bind at the active site.

WORD ROOTS

end- = inside (e.g., an *end*ergonic reaction requires an input of energy in order to proceed)

exo- = outside (e.g., *exo*cytosis is the release of material from the cell)

thermo- = heat (e.g., a *thermo*meter is a tool for measuring temperature)

KEY TERMS

activation energy _____

active site _____

adenosine triphosphate (ATP) _____

allosteric regulation _____

catalyst _____

coenzyme _____

competitive inhibition _____

coupled reaction _____

endergonic reaction _____

energy _____

enzyme _____

exergonic reaction _____

first law of thermodynamics _____

kinetic energy _____

metabolic pathway _____

metabolism _____

potential energy _____

second law of thermodynamics _____

substrate _____

thermodynamics _____

FLASH CARDS

To use the flash cards, tear the page from the book and cut along the dashed lines. The key term appears on one side of the flash card, and its definition appears on the opposite side.

activation energy	coupled reaction
active site	endergonic reaction
adenosine triphosphate (ATP)	energy
allosteric regulation	enzyme
catalyst	exergonic reaction
coenzyme	first law of thermodynamics
competitive inhibition	kinetic energy

a chemical reaction in which an endergonic reaction is powered by an exergonic reaction; all "uphill" or endergonic reactions require an input of energy; this energy is supplied by "downhill" or exergonic reactions

the energy required to initiate a chemical reaction; enzymes lower the activation energy of a reaction, thereby greatly speeding up the rate of the reaction

a chemical reaction in which the ending set of molecules (the products) contains more energy than the starting set of molecules (the reactants)

the portion of an enzyme that binds with a substrate, thus helping transform it

the capacity to bring about movement against an opposing force

a nucleotide that serves as the most important energy-transfer molecule in living things; ATP powers a broad range of chemical reactions by donating one of its three phosphate groups to these reactions; in the process, it becomes adenosine diphosphate (ADP), which reverts to being ATP when a third phosphate is added to it

a chemically active protein that speeds up, or in practical terms enables, chemical reactions in living things

the regulation of an enzyme's activity by means of a molecule binding to a site on the enzyme other than its active site

a chemical reaction in which the starting set of molecules (the reactants) contain more energy than the final set of molecules (the products)

a substance that retains its original chemical composition while bringing about a change in a substrate; enzymes are catalysts in chemical reactions; one enzyme can carry out hundreds or thousands of chemical transformations without itself being transformed

energy cannot be created or destroyed, but can only be transformed from one form to another

a type of accessory molecule that binds to the active site of an enzyme, thus allowing the enzyme to bind to its substrate; many vitamins are important coenzymes

energy in motion; a rolling rock has kinetic energy

a reduction in the activity of an enzyme by means of a compound other than the enzyme's usual substrate binding with the enzyme in its active site

metabolic pathway

metabolism

potential energy

second law of thermodynamics

substrate

thermodynamics

energy transfer always results in a greater amount of disorder (entropy) in the universe

a sequential set of enzymatically controlled reactions in which the product of one reaction serves as the substrate for the next

a substance whose chemical alteration is facilitated by an enzyme

the sum of all chemical reactions carried out by a cell or larger organism

the study of energy

stored energy; a rock perched at the top of a hill has potential energy

SELF TEST

Once you have finished studying this chapter, close your books, grab a pencil, and spend the next 15 to 20 minutes completing this practice test.

Compare and Contrast

For each of the following paired terms, write a sentence of comparison ("Both") and a sentence of contrast ("However,").

endergonic/exergonic
enzyme/coenzyme
potential energy/kinetic energy
competitive inhibitor/allosteric inhibitor
metabolic pathway/metabolism

Short Answer

1. Why do we need to supply our bodies with new energy sources on a fairly regular basis?

2. Define *kinetic energy*.

3. How can coupled reactions be used to make more-ordered structures, such as starch, from less-ordered molecules, such as glucose?

4. Why is ATP essential to life?

5. What is meant by the term *metabolic pathway*?

6. What are coenzymes? What do they do?

7. How does allosteric regulation affect susbtrate-enzyme binding?

Multiple Choice

Circle the letter that best answers the question.

1. In which of the following reactions would energy be created?
   a. enzyme
   b. entropy
   c. endergonic
   d. enzymatic
   e. none of the above

2. Imagine that you attempt to pour sand from one cup to another. Which of the following terms best describes the sand that spills instead of landing in the second cup?
   a. exergonic
   b. endergonic
   c. entropy
   d. exogenous
   e. exercise

3. Which of the following terms would be used to describe the process by which plants convert the energy of sunlight into the energy of chemical bonds in food?
   a. entropy
   b. endergonic
   c. exergonic
   d. glycolysis
   e. substrate level phosphorylation

4. Which of the following terms correctly refers to ATP?
   a. adenosyl transphosphate
   b. andesine triphosphate
   c. adenosyl tripolymer
   d. adenosine triphosphate
   e. adenylyl tricyclase

5. A coupled reaction:
   a. involves a male enzyme as well as a female one.
   b. is the result of mixing acids and bases.
   c. is a set of exergonic and endergonic reactions.
   d. means that an enzyme is paired with its coenzyme during the reaction.
   e. involves the transfer of oxygen from one substrate to another.

6. While serving as a visiting professor on the planet Freedonia, you discover a life-form that does not use ATP as its primary energy transfer molecule. Instead, the organism uses michelerene phosphate (MiP). Michelerene phosphate stores energy by adding positively charged michelerene groups to the backbone phosphate. Which of the following would have the highest energy state (the greatest potential energy)?
   a. MiP+
   b. MiP++
   c. MiP+++
   d. MiP−
   e. MiP−−

7. The process of converting MiP++ to MiP+++ (see question 6) is:
   a. exergonic.
   b. endergonic.
   c. potentiation.
   d. conservative.
   e. endoscopic.

8. How does the cell use ATP to drive uphill reactions?
   a. by using energy to both make ATP and drive the uphill reaction
   b. by converting ATP to ADP, which releases energy to drive the uphill reaction

c. by converting ADP to ATP, which releases energy to drive the uphill reaction

d. by using enzymes that make the uphill reaction become downhill

e. a and c

9. Why is ATP such a good energy currency molecule?
   a. It is very stable.
   b. Its negatively charged phosphate groups repel one another.
   c. It has three positive charges that repel one another, forming an unstable molecule.
   d. It has double covalent bonds, which store twice as much energy.
   e. It is a large molecule.

10. What part of a reaction is changed by an enzyme?
    a. the energy of the reactants
    b. the activation energy
    c. the energy of the products
    d. the energy difference between the products and the reactants
    e. the alteration energy

11. At which site does an allosteric inhibitor bind?
    a. the active site of an enzyme
    b. the active site on the products
    c. the active site on the reactants
    d. a site on the enzyme away from the active site
    e. a site on the reactants not affected by the chemical reaction

12. The enzyme lactase participates in what reaction?
    a. the breakdown of lactose
    b. the formation of fructose
    c. the breakdown of glucose
    d. the formation of protein
    e. the formation of lipids

13. A number of diseases, including rickets, beriberi, and scurvy, have been linked to vitamin deficiencies. How could these deficiencies cause illness?
    a. by decreasing cellular ADP
    b. by decreasing the efficiency of cellular enzymes
    c. by increasing the reproduction of bacteria that produce the vitamins
    d. by causing the cellular nucleus to dissolve
    e. by interfering with protein synthesis

14. How do living things create energy?
    a. They can't; the first law of thermodynamics states that energy can neither be created nor destroyed.
    b. They can't; the second law of thermodynamics states that energy can neither be created nor destroyed.

c. by using sunlight

d. Chemical reactions within cells create energy during phosphate transfer.

e. by breaking down carbohydrates like glucose

15. Almost all the living organisms on the planet use the _____ as the ultimate source of energy.
    a. chemical reactions within cells
    b. chemical reactions between cells and their environment
    c. sun
    d. ocean currents
    e. oxygen gradient

16. Which of the following conditions represents potential energy?
    i. table sugar ready to ingest
    ii. a diver about to dive off the 10-m platform at the Olympics
    iii. a gerbil running on its exercise wheel in its cage
    a. i only
    b. ii only
    c. i, ii, and iii
    d. i and ii
    e. ii and iii

17. Which of the following is *not* a type of work exhibited by living organisms?
    a. mechanical
    b. transport
    c. synthetic
    d. None of these is a type of work exhibited by living organisms,
    e. All of the given types of work are found in living organisms.

18. Which of the following is an example of synthetic work?
    a. building a carbohydrate out of simple sugars
    b. breaking down proteins into amino acids
    c. contracting a muscle to move your arm through space
    d. transporting materials across a cell membrane
    e. breaking down carbohydrates into glucose molecules

19. For the following reaction, determine whether it is endergonic or exergonic.

    Substance A + Substance B + ATP → Substance C

    a. exergonic
    b. endergonic
    c. The reaction is both endergonic and exergonic.
    d. The reaction is ergonically neutral.
    e. The reaction is either ergonically neutral or endergonic.

20. The synthesis of ATP from ADP and inorganic phosphate is:
  a. exergonic.
  b. endergonic.
  c. both endergonic and exergonic.
  d. ergonically neutral.
  e. either ergonically neutral or endergonic.

21. Your research mentor has asked you to order more of the enzyme you have been working with in the lab. Your project involves measuring the amount of glucose found in the celluose produced by different species. Which of the following should you order based on the substrate you are using?
  a. phospholipase
  b. DNAase
  c. cellulase
  d. glucose
  e. cellulose

22. Which form of enzyme regulation involves a change of the enzyme's shape?
  a. competitive inhibition
  b. allosteric regulation
  c. aminotropic inhibition
  d. homosteric regulation
  e. tyrosine cross-linkage

Use the figure below to answer the questions that follow:

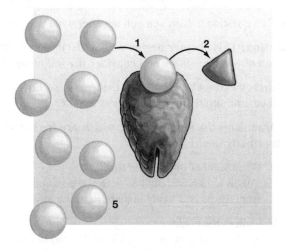

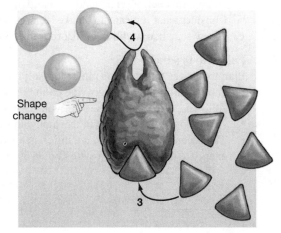

23. Which site is the active site?
  a. 1
  b. 2
  c. 3
  d. 4
  e. 5

24. Which molecule is an allosteric inhibitor?
  a. 1
  b. 2
  c. 3
  d. 4
  e. 5

25. What types of macromolecules are enzymes?
  a. protein
  b. lipid
  c. carbohydrate
  d. nucleic acid
  e. steroids

WHAT'S IT ALL ABOUT?

Here's a question to help you pull together what you've learned so far using this text.

Question: This chapter begins with a discussion of the first and second laws of thermodynamics. Reread the section about these laws, and consider the following observations:

1. **Energy cannot be created or destroyed, so how can organisms make the energy they need to move, reproduce, and interact with the environment?**

2. **Organisms are made of cells, which are highly ordered structures. Is this a violation of the second law of thermodynamics?**

   1. **What kind of question is this?**
      This is a variation on the "defend a position" question. You are being asked to explain why two observations from the natural world that appear to be in violation of the laws of thermodynamics are truly not.

   2. **Collect the evidence.**
      a. For your benefit, write down the two laws of thermodynamics in question, and consider what it means to "transform energy" and "tend to disorder."
      b. Consider what it means to "make" energy in a cell—is it really creating energy, or transforming it?
      c. Likewise, when we think of disorder in a "system," what defines a system?

   3. **Pull it all together.**
      Plan to answer the questions in order, so you will need two paragraphs. Start with summarizing the laws of thermodynamics, and then provide examples of how living organisms do not violate these laws. You can use the following topic sentences to get you started.
      • According to the first law of thermodynamics, energy cannot be created or destroyed, only transformed.
      • The second law of thermodynamics states that the universe tends to become more disordered during these energy transformations, a state known as an increase in entropy.

CHAPTER 7 VITAL HARVEST: DERIVING ENERGY FROM FOOD

Basic Chapter Concepts

- ATP is generated by the fall of energized electrons through various reduction-oxidation reactions.
- Three metabolic pathways extract energy from food in complex organisms—glycolysis, the Krebs cycle, and the electron transport chain.
- Breaking the bonds of ATP supplies energy for metabolic reactions.

CHAPTER SUMMARY

7.1 Energizing ATP: Adding a Phosphate Group to ADP
- The molecular structure of ATP contains three phosphate groups. When the bonds between the phosphate groups are broken, small amounts of energy are released to power cellular reactions.

7.2 Electrons Fall Down the Energy Hill to Drive the Uphill Production of ATP
- Electrons are harvested from glucose, a high-energy molecule.
- Oxidation-reduction (redox) reactions always happen in pairs; molecules that attract electrons strongly accept electrons from high-energy molecules. The molecule that accepts the electrons becomes reduced while the high-energy donor becomes oxidized.

7.3 The Three Stages of Cellular Respiration: Glycolysis, the Krebs Cycle, and the Electron Transport Chain
- The Krebs cycle and the electron transport chain (ETC) are evolutionarily more recent additions to the energy-harvesting machinery. These two stages require oxygen and produce an enormous amount of energy (32 ATP molecules) from a single glucose molecule.
- These reactions occur within the mitochondria, the powerhouse of the cell.

7.4 First Stage of Respiration: Glycolysis
- Glycolysis is the obligatory first stage in energy harvesting; it exists in all organisms.
- Glycolysis is very efficient, producing only 2 ATP molecules for each glucose molecule oxidized. It provides essential substrates for the Krebs cycle and the electron transport chain—2 NADH and 2 pyruvic acid molecules per glucose molecule oxidized.

7.5 Second Stage of Respiration: The Krebs Cycle
- The Krebs cycle, also called the citric acid cycle, accepts the carbons from pyruvic acid as the high-energy intermediate called acetyl CoA, and it oxidizes them completely to carbon dioxide, which we exhale.
- The Krebs cycle "saves" the energy of oxidation reactions in the form of the reduced electron carrier molecule NADH.

7.6 Third Stage of Respiration: The Electron Transport Chain
- The ETC allows the high-energy NADH molecule to donate its electrons to other electron carriers that use the energy released in this downhill transfer of electrons to pump protons up an energy hill against their concentration gradient. Releasing these protons to flow down their concentration gradient releases the energy needed to make ATP.
- Electron transfer proceeds down the energy hill; that is, the process starts with a high-energy molecule donating electrons to a lower-energy acceptor molecule. The molecule donating electrons becomes oxidized, whereas the acceptor becomes reduced. Oxygen is not required for redox reactions; any molecule that donates electrons can be oxidized, and any molecule that can accept electrons can be reduced.

7.7   Other Foods, Other Respiratory Pathways
- Organisms use different types of molecules as energy sources.
- Molecules other than glucose can be oxidized to produce energy, but all of these alternative oxidation pathways eventually feed into glycolysis, the Krebs cycle, or the ETC.

WORD ROOTS

glyco- = relating to sugar; **lysis** = decomposition (e.g., *glycolysis* is the metabolic breakdown [decomposition] of glucose by cells)

KEY TERMS

ATP synthase _____

cellular respiration _____

citric acid cycle _____

electron carrier _____

electron transport chain (ETC) _____

glycolysis _____
_____

Krebs cycle _____
_____

nicotinamide adenine dinucleotide (NAD) _____
_____

oxidation _____
_____

redox reaction _____
_____

reduction _____
_____

FLASH CARDS

To use the flash cards, tear the page from the book and cut along the dashed lines. The key term appears on one side of the flash card, and its definition appears on the opposite side.

ATP synthase

Krebs cycle

cellular respiration

nicotinamide adenine dinucleotide (NAD)

citric acid cycle

oxidation

electron carrier

redox reaction

electron transport chain (ETC)

reduction

glycolysis

the second stage of cellular respiration, occurring in the inner compartment of mitochondria; the Krebs cycle is the major source of electrons that power the third stage of respiration, the electron transport chain; also known as the citric acid cycle

an enzyme functioning in cellular respiration that brings together ADP and inorganic phosphate molecules to produce ATP

the most important intermediate electron carrier in cellular respiration

the three-stage, oxygen-dependent harvesting of energy that goes on in most cells; the three stages are glycolysis, the Krebs cycle, and the electron transport chain

loss of one or more electrons by an atom or a molecule; important in energy transfer in living things as part of redox reactions, in which one substance undergoes reduction (a gain in electrons) by oxidizing another

another name for the Krebs cycle, one of the three main sets of steps in cellular respiration; named for the first product of the cycle, citric acid

a combination of a reduction and an oxidation reaction in which the electrons lost from one substance in oxidation are gained by another in reduction

a molecule that serves to transfer electrons from one molecule to another in ATP formation; the most important electron carrier in ATP formation is NAD^+ (NADH in its reduced form); the other major carrier is FAD ($FADH_2$)

gain of one or more electrons by an atom or a molecule; important in energy transfer in living things as part of redox reactions, in which one substance is reduced by oxidizing (or removing electrons from) another substance

the third stage of cellular respiration, occurring within the inner membrane of the mitochondria, in which most of the ATP is formed

the first stage of cellular respiration, occurring in the cytosol; for some organisms, glycolysis is the sole means of extracting energy from food; in most organisms, it is a means of extracting some energy and a necessary precursor to the other two stages of cellular respiration, the Krebs cycle and the electron transport chain

SELF TEST

Once you have finished studying this chapter, close your books, grab a pencil, and spend the next 15 to 20 minutes completing this practice test.

Compare and Contrast

For each of the following paired terms, write a sentence of comparison ("Both") and a sentence of contrast ("However,").

oxidation/reduction
mitochondria/cytoplasm
pyruvic acid/citric acid
cellular respiration/glycolysis
electron transport chain/NAD

Short Answer

1. Why is the production of ATP a process that requires energy?

2. Explain the relationship between the terms *redox* and *coupled reactions*.

3. If you were building an artificial cell, what important parts must it have to be highly efficient at producing energy?

4. Glycolysis is an inefficient process, but some cells manage to live on it. How does this occur?

5. What is the evolutionary significance of the Krebs cycle?

6. The ETC transports electrons and pumps protons. How are these events related to ATP synthesis?

7. You have decided to go on the Atkins diet. Describe how the pathways to glycolysis will differ.

Multiple Choice

Circle the letter that best answers the question.

1. Why are ions pumped across the inner mitochondrial membrane?
   a. So they can get to the other side.
   b. They are the waste products of cellular respiration.
   c. Their movement back across the membrane generates ATP.
   d. It is the step that makes the Krebs cycle work as a cycle.
   e. It generates acetyl CoA.

2. Why is the Krebs cycle a cycle?
   a. because the most important molecule, oxaloacetate, is circular
   b. because whereas the rest of cellular respiration happens during the day, the Krebs cycle takes place at night
   c. because the first molecule in the pathway is also the last
   d. because it takes place in the mitochondria, which are round
   e. because that is the name that Krebs chose

3. Which stage yields the greatest amount of ATP?
   a. Krebs cycle
   b. acetyl CoA formation
   c. electron transport chain
   d. glycolysis
   e. oxidation

4. Which stage is evolutionarily the oldest?
   a. Krebs cycle
   b. acetyl CoA formation
   c. electron transport chain
   d. glycolysis
   e. oxidation

5. Which stage takes place in the cytoplasm?
   a. Krebs cycle
   b. ATP synthase
   c. electron transport chain
   d. glycolysis
   e. oxidation

6. Which stage releases the first molecule of carbon dioxide?
   a. Krebs cycle
   b. acetyl CoA formation
   c. electron transport chain
   d. glycolysis
   e. oxidation

7. Which of the following best describes an oxidation reaction?
   a. a substance gains an oxygen atom
   b. a substance gains electrons
   c. a substance gives an oxygen atom to another substance
   d. a substance donates electrons to another substance
   e. a substance usually combusts

8. What is the correct order of the stages of cellular respiration?
   a. Krebs cycle, electron transport chain, glycolysis
   b. electron transport chain, Krebs cycle, glycolysis
   c. glycolysis, Krebs cycle, electron transport chain
   d. glycolysis, electron transport chain, Krebs cycle
   e. Krebs cycle, glycolysis, electron transport chain

9. Which molecules can be oxidized to generate ATP?
   a. glucose
   b. fats
   c. proteins
   d. glycogen
   e. all of the above

10. Back on the planet Freedonia, with infinite resources available to you, you decide to try implanting mitochondria into bacterial cells. Assuming your experiment is successful, what results would you predict?
    a. no change
    b. an increase in ATP yield/glucose molecule
    c. no change in oxygen usage by the bacteria
    d. a decrease in oxygen usage by the bacteria
    e. the bacteria will begin to produce alcohol

11. Suppose you were to then supply the cells with the means of making acetyl CoA. What would you expect to see?
    a. no change
    b. an increase in ATP yield/glucose molecule
    c. no change in oxygen usage by the bacteria
    d. a decrease in oxygen usage by the bacteria
    e. the bacteria will begin to produce alcohol

12. You and your brother are racing each other home. The first one home will get the last piece of your mom's delectable chocolate cake. As you run, you feel a burning sensation in your leg muscles, which you know is caused by lactic acid fermentation. Why do your muscle cells switch from aerobic respiration to anaerobic respiration, which generates fewer ATP per glucose molecule used?
    a. Aerobic respiration is expensive, and periodically the cells "take a break" and use anaerobic respiration for a period of time.
    b. Muscle cells don't have mitochondria and therefore are incapable of performing aerobic respiration.
    c. Aerobic respiration takes longer to get started, so muscle cells switch to fermentation for short, intense activities.
    d. Anaerobic respiration can be carried out using muscle proteins (actin and myosin) as an energy source, so the muscles can perform longer.
    e. Some muscle cells have more mitochondria than others, and the ones with more mitochondria cause the burning sensation.

13. When proteins are used as the major energy source of cellular respiration:
    a. they are transformed into glucose.
    b. they are transformed into lipids.
    c. the component amino acids are shunted into glycolysis.
    d. the energy produced is of a lower quality than glucose.
    e. the energy produced is of a higher quality than glucose.

14. In an experiment, rats were given glucose that contained a small amount of radioactive oxygen. The mice were closely watched, and over a short time period radioactive oxygen atoms were found in:
    a. carbon dioxide.
    b. water.
    c. ATP.
    d. oxygen.
    e. NADH.

15. A drug that blocks the enzymes that control the electron transport chain would probably cause the organism to:
    a. starve to death.
    b. get fatter because of the extra glucose.
    c. find another cellular pathway to produce energy.
    d. have a slight decrease in activity level.
    e. none of the above

16. George lacks the metabolic pathways to produce glycogen and must snack constantly in order to live. This is because:
    a. glycogen is necessary for life.
    b. without stored glycogen, George must have a constant glucose intake.
    c. he is following a special diet.
    d. he can live only on glucose.
    e. his cells have now switched to a different energy source.

17. Just as you must have money in the bank to earn interest on that money, cells must:
    a. use two ATP molecules to start the initial phase of glycolysis.
    b. use ATP molecules to start the Krebs cycle.
    c. take away one ATP molecule from glucose.
    d. take away two ATP molecules from glucose.
    e. use three ATP molecules to start the initial phase of glycolysis.

18. The main source of energy for humans is:
    a. fats.
    b. carbohydrates.
    c. proteins.
    d. nucleotides.
    e. steroids.

Match the following terms with their description. Each choice may be used once, more than once, or not at all.
    a. glycolysis
    b. electron transport chain
    c. acetyl CoA formation
    d. Krebs cycle
    e. oxidation
    f. reduction

19. _____ Process by which an electron is lost

20. _____ A three-carbon molecule is converted to a two-carbon molecule; a molecule of carbon dioxide is released

21. _____ Process by which an electron is gained

22. _____ Stage at which oxygen is required

23. _____ Takes place in inner mitochondrial compartment

24. _____ Shared by all living organisms

25. _____ Oldest form of energy production

WHAT'S IT ALL ABOUT?

Here's a question to help you pull together what you've learned so far using this text.

Question: Every November, millions of Americans sit down to a feast, probably the largest single meal they will eat all year. After this gorge, most of us leave the table to watch televised football or fall asleep. Based on your new knowledge of cellular respiration, and your current knowledge of the structures of biomolecules, describe how our bodies use the turkey, dressing, vegetables, and pumpkin pie we have just eaten.

1. **What kind of question is this?**

   You are being asked to apply your knowledge to solve a "problem," that is, to figure out an explanation for a novel scenario.

2. **Collect the evidence.**

   You already know from past experience the major food groups correspond to our major classes of biomolecules—proteins, carbohydrates, and fats. Using what you have learned about deriving energy from food, describe the fate of the molecules consumed during Thanksgiving dinner.

3. **Pull it all together.**

   The question has given you two key points while describing this scenario: the quantity of food consumed (large) and the level of activity following this meal (low). You must consider both of these points in framing your answer. You may use the following topic sentence to get started:

   Our Thanksgiving feast is made up of various biomolecules: protein and fat in the turkey, mostly carbohydrates in the dressing and vegetables, and mostly fats and simple sugars in the pie.

CHAPTER 8 THE GREEN WORLD'S GIFT: PHOTOSYNTHESIS

Basic Chapter Concepts

- Photosynthesis captures the light energy of the sun and stores it in the carbohydrates that plants produce. Because most organisms depend on plants as a source of fuel (directly or indirectly), photosynthesis makes life on Earth possible.
- Photosynthesis has two stages—the first saves light energy as energetic electrons in NADPH and releases oxygen, and the second delivers the energy in the electrons of NADPH to a three-carbon sugar to make food.
- Alternative metabolic cycles make photosynthesis possible under difficult environmental conditions.

CHAPTER SUMMARY

8.1 Photosynthesis and Energy
- Photosynthesis creates the food for all organisms from sunlight, water, and carbon dioxide. The food is broken down in the energy-harvesting reactions of glycolysis, the Krebs cycle, and the electron transport chain (cellular respiration).
- Photosynthesis generates oxygen as a by-product.
- Photosynthesis and cellular respiration are opposing reactions; photosynthesis is an endergonic process that stores solar energy as chemical energy in food, whereas cellular respiration is an exergonic process that releases chemical energy for mechanical or chemical work.

8.2 The Components of Photosynthesis
- The photo stage of photosynthesis occurs within the membrane networks of a specialized cellular organelle, the chloroplast. Chloroplasts contain pigments that allow the capture of solar energy and the conversion of that energy into chemical energy through a series of redox reactions.
- The second stage is the synthesis stage, also known as the Calvin cycle. In the Calvin cycle, the products of the light reactions and carbon are used to build organic molecules.

8.3 Stage 1: The Steps of the Light Reactions
- The working units of photosynthesis are two photosystems. These photosystems transfer electrons taken from water and are energized by absorbing solar energy through a series of redox reactions, converting that solar energy into reducing power stored as NADPH and into a proton gradient that can make ATP.

8.4 What Makes the Light Reactions So Important?
- During the light reactions, energized electrons are used to generate ATP. As these electrons leave their reaction center, they are replaced by others. These electrons are released by the splitting of water molecules, a reaction that also generates oxygen.

8.5 Stage 2: The Calvin Cycle
- The working unit of carbohydrate synthesis is the enzyme rubisco. Rubisco fixes carbon from atmospheric CO_2 as 3-phosphoglyceric acid, a three-carbon sugar that can be used to make glucose.
- The substrate for rubisco is RuBP, a five-carbon sugar.

8.6 Photorespiration and the C_4 Pathway
- In photosynthesis, there is a "glitch" known as photorespiration. Photorespiration occurs because rubisco can bind to oxygen as well as carbon dioxide; however, when it binds to oxygen, energy is used, but no sugars are made.
- Alternative pathways have developed in plants in warm, in wet and warm, and in dry climates to get around this wasteful process.
- C_4 plants fix carbon using a different reaction and then release carbon dioxide into the bundle-sheath cells for use in the Calvin cycle, at the cost of ATP energy.

8.7   Another Photosynthetic Variation: CAM Plants

- CAM plants—common in hot, dry climates—fix carbon using the CAM pathway at night and wait until light is available before proceeding with the Calvin cycle. The CAM pathway uses an acid intermediate to hold or fix carbon until it is required by the plant.

WORD ROOTS

photo- = light (e.g., *photo*synthesis is the process by which organic molecules are built using energy from sunlight)

chloro- = chlorophyll (e.g., the *chloro*plast is the photosynthetic organelle in plants and contains chlorophyll)

KEY TERMS

C_4 photosynthesis _____

Calvin cycle _____

CAM photosynthesis _____

chlorophyll *a* _____

chloroplast _____

fixation _____

photorespiration _____

photosynthesis _____

photosystem _____

reaction center _____

rubisco _____

stomata _____

stroma _____

thylakoid _____

FLASH CARDS

To use the flash cards, tear the page from the book and cut along the dashed lines. The key term appears on one side of the flash card, and its definition appears on the opposite side.

C_4 photosynthesis	photosynthesis
Calvin cycle	photosystem
CAM photosynthesis	reaction center
chlorophyll *a*	rubisco
chloroplast	stomata
fixation	stroma
photorespiration	thylakoid

the process by which certain groups of organisms capture energy from sunlight and convert this solar energy into chemical energy that is initially stored in a carbohydrate

a form of photosynthesis in which carbon dioxide is first fixed into a four-carbon molecule and then transferred to special bundle-sheath cells in which the Calvin cycle is undertaken; used primarily by plants in warm-weather environments to reduce the photorespiration that undercuts carbohydrate production in C_3 plants

an organized complex of molecules within a thylakoid membrane that, in photosynthesis, collects solar energy and transforms it into chemical energy

the set of steps in photosynthesis in which an energy-rich sugar is produced by means of two essential processes—the fixing of atmospheric carbon dioxide into a sugar and the energizing of this sugar with the addition of electrons supplied by the light-dependent reactions of photosynthesis

a molecular complex in a chloroplast that, in photosynthesis, transforms solar energy into chemical energy

a form of photosynthesis undertaken by some plants in hot, dry climates, in which carbon fixation takes place at night and the Calvin cycle during the day; CAM metabolism allows plants to preserve water by opening their stomata only at night

an enzyme that allows plants to incorporate atmospheric carbon dioxide into their own sugars during the process of photosynthesis

the primary pigment of the chloroplast, found embedded in its membranes; together with the accessory pigments, chlorophyll *a* absorbs some wavelengths of sunlight in the first step of photosynthesis

microscopic pores, found in greatest abundance on the undersides of leaves, that allow plants to exchange gases with the atmosphere; carbon dioxide moves into plants through the stomata, while oxygen and water vapor move out

the organelle within plant and algae cells that is the site of photosynthesis

in plants and algae, the liquid material of chloroplasts that is the site of the Calvin cycle

the process of incorporating a gas into an organic molecule; in photosynthesis, atmospheric carbon dioxide (CO_2) is fixed into the sugars of photosynthesizing organisms, such as plants; in ecology, atmospheric nitrogen (N_2) is fixed into ammonia (NH_3) by certain bacteria

a flattened, membrane-bound sac in the interior of a chloroplast that serves as the site for the light reactions in photosynthesis

a process in which the enzyme rubisco undercuts carbon fixation in photosynthesis by binding with oxygen instead of with carbon dioxide; this wasteful reaction takes place most frequently in C_3 plants; the warm-weather C_4 plants have evolved a means of reducing photorespiration

SELF TEST

Once you have finished studying this chapter, close your books, grab a pencil, and spend the next 15 to 20 minutes completing this practice test.

Compare and Contrast

For each of the following paired terms, write a sentence of comparison ("Both") and a sentence of contrast ("However,").

chloroplast/chlorophyll
photorespiration/photosynthesis
stroma/grana
reaction center/rubisco
Calvin cycle/CAM photosynthesis

Short Answer

1. Photosynthesis makes life on Earth possible. Why?

2. What is the source of energy that drives photosynthesis? Be specific.

3. What is a reaction center? What stage of photosynthesis requires it?

4. As the result of the light reactions, plants release oxygen. What process generates oxygen?

5. Explain the significance of RuBP.

6. What is photorespiration? What atmospheric conditions make photorespiration more likely to occur?

7. Explain the major differences between the CAM and C_4 adapations to photorespiration.

Multiple Choice

Circle the letter that best answers the question.

1. Why is the C_4 pathway called C_4?
   a. because it involves four steps
   b. because it initially fixes into a four-carbon molecule
   c. because only four types of plants have this pathway
   d. because it involves the four C's (cut, clarity, . . .)
   e. because four molecules of CO_2 are fixed into carbohydrate

2. What is the initial electron donor in photosynthesis?
   a. O_2
   b. rubisco
   c. NADPH
   d. H_2O
   e. PSI

3. What are the products of photosynthesis?
   a. plant material, H_2O, and NADP+
   b. CO_2 and plant material
   c. CO_2 and O_2
   d. O_2 and plant material
   e. minerals, CO_2, and H_2O

4. Which of the following is (are) true about the relationship between the Calvin and Krebs cycles?
   a. One uses glucose, while the other produces it.
   b. Both are named for the same man.
   c. Both are involved in energy transformation.
   d. Both produce ATP.
   e. a and c

5. Which of the following terms does *not* refer to light-dependent reactions?
   a. photosystem I
   b. photosystem II
   c. NADPH
   d. RuBP
   e. chlorophyll

6. Which of the following is equal to one-half of a glucose molecule?
   a. ATP
   b. NADPH
   c. G_3P
   d. STP
   e. ABC

7. Stroma is:
   a. the collection of molecules that light strikes first.
   b. the organelle where photosynthesis takes place.
   c. the opening in leaves that allows the passage of water and carbon dioxide.
   d. the liquid found inside the chloroplast.
   e. the nitrogenous residue of the Calvin cycle.

8. For the Calvin cycle to begin, RuBP must combine with _____ and rubisco.
   a. water
   b. ATP
   c. sunlight
   d. carbon dioxide
   e. ruminants

9. Which of the following is the process that ultimately feeds the world?
   a. decomposition
   b. photosynthesis
   c. respiration
   d. reproduction
   e. regeneration

10. As you finish up your sabbatical on the planet Freedonia, you hear of a type of photosynthetic plant that cannot make rubisco and therefore cannot complete the Calvin cycle. You correctly think to yourself:
    a. "They must not have chloroplasts."
    b. "They can't really exist without the ATP produced in the Calvin cycle."
    c. "They are probably less complex, unicellular organisms such as bacteria, because they lack energy-storage capability."
    d. "They are likely to be very large, multicellular organisms such as sequoia trees because of their rapid ATP turnover."
    e. "Plants on Earth don't need rubisco either."

11. Which of the following might be used as proof that global warming is occurring?
    a. a decrease in photorespiration rates
    b. an increase in the number of species with C_4 adaptations
    c. a darkening in color of photosynthetic pigments
    d. a decrease in the cooking time required for fresh-picked vegetables
    e. a decrease in the number of hurricanes in the Atlantic ocean

12. Which of the following is *not* found in both plant and animal cells?
    a. ATP
    b. G_3P
    c. mitochondria
    d. glucose
    e. all of the above

13. What is photorespiration?
    a. when rubisco binds oxygen and no growth occurs
    b. cellular respiration in plant cells
    c. synthesis of carbohydrates from nitrogen-rich sources
    d. cellular respiration in chloroplasts
    e. the light-dependent reactions

14. What do cell respiration and photosynthesis have in common?
    a. both produce glucose
    b. both produce (and use) ATP
    c. both take place in animal cells
    d. both require sunlight
    e. both use glucose

15. Which plant organelles convert the energy in sunlight into carbohydrates?
    a. endoplasmic reticulum
    b. nuclei
    c. mitochondria
    d. chloroplasts
    e. lysosomes

16. You find an arctic tundra plant that appears to use the C_4 pathway. Why is this a surprising finding?
    a. C_4 metabolism is found exclusively in water plants.
    b. C_4 metabolism is found only in animals.
    c. C_4 metabolism is found in warm-climate plants.
    d. C_4 metabolism does not exist.
    e. C_4 metabolism uses compounds not found in the tundra.

17. How does the CAM pathway differ from the C_4 pathway?
    a. CAM photosynthesis produces fructose instead of glucose.
    b. C_4 photosynthesis does not require sunlight.
    c. CAM photosynthesis requires opening stomata only at night.
    d. C_4 photosynthesis is found only in cooler climates.
    e. CAM is found in the chloroplast, whereas C_4 takes place in the mitochondria.

18. The reaction center:
    a. stores ATP.
    b. transforms light energy into chemical energy.
    c. produces glucose from G_3P.
    d. is made up of rubisco molecules.
    e. is the site of photorespiration.

19. What pathway in photosynthesis fixes carbon?
    a. the electron transport chain
    b. the light reactions
    c. the Calvin cycle
    d. photorespiration
    e. c and d

20. Where do the electrons from photosystem II end up?
    a. H_2O
    b. photosystem I
    c. NADPH
    d. O_2
    e. CO_2

21. Reactions that involve both the gain and loss of electrons are described as _____ reactions.
    a. endergonic
    b. exergonic
    c. redox
    d. cycle
    e. conservation

22. What is rubisco?
    a. a carbohydrate
    b. a lipid
    c. a chlorophyll
    d. an enzyme
    e. a coenzyme (vitamin)

23. What provides the energy to boost an electron up the energy hill from photosystem II?
    a. sunlight
    b. ATP
    c. glucose metabolism
    d. splitting H_2O
    e. sunlight

24. Compare the electron transfer process of photosynthesis with the electron transfer chain of mitochondria. Which of the following are similarities?
    a. both use electrons to generate ATP
    b. both use electron carriers to donate electrons
    c. both are used in transforming energy
    d. both are found in eukaryotes
    e. all of the above

25. The advantage of CAM photosynthesis over C_3 and C_4 photosynthesis is that:
    a. CAM has less water loss.
    b. CAM has more efficent ATP usage.
    c. CAM has less photorespiraton.
    d. CAM requires less CO_2 to produce glucose.
    e. CAM is found in all climates.

WHAT'S IT ALL ABOUT?

Here's a question to help you pull together what you've learned so far using this text.

Question: In this chapter, we have explored how plants make carbohydrates (their food and ours) from nothing more than water, carbon dioxide, and the energy of sunlight. In the previous chapter, we saw how we (and plants) can extract this energy in the process of cellular respiration. Despite the differences in the function of these two pathways, they share a remarkable number of features. Describe how photosynthesis differs from cellular respiration and how the two pathways are the same.

1. **What kind of question is this?**
   This is a "compare and contrast" question.

2. **Collect the evidence.**
   Make a two-column table. List the similarities in one column, and the differences in the other. Use the information in your textbook to collect your evidence.

3. **Pull it all together.**
   Because the question phrases the problem as "despite the differences . . . share features," we would choose to point out the differences first and then provide examples of the similarities. On the surface, photosynthesis and cellular respiration are opposing pathways. You might use a starting sentence such as, "On the surface, photosynthesis and cellular respiration are opposing pathways" to introduce the differences, and then might follow up with a sentence beginning with "Despite these differences," or "However, both pathways share . . . ."

CHAPTER 9 GENETICS AND CELL DIVISION

Basic Chapter Concepts

- DNA encodes all of the information necessary for an organism to develop, grow, repair itself, and duplicate itself.
- Information is stored within a gene, the sequence of nitrogen-containing bases. Each gene specifies a different protein responsible for executing a cellular function.
- Genetic information is duplicated and transferred during the cell cycle, a multistage process that involves copying all of the genes (interphase) and equal division of chromosomes (mitosis), followed by division of the cell into two daughters (cytokinesis).

CHAPTER SUMMARY

9.1 An Introduction to Genetics
- DNA contains instructions for producing protein.
- Genetics is concerned with the storage, duplication, and transfer of information.
- An organism's genome consists of its entire collection of genes.

9.2 An Introduction to Cell Division
- Each time a cell replicates, its entire genome must be copied.
- Cell division involves duplication of DNA and the splitting of the parent cell into two daughter cells.

9.3 DNA Is Packaged in Chromosomes
- Different organisms have different numbers of chromosomes.
- Chromosomes consist of tightly condensed DNA that is wound around proteins.
- In humans, chromosomes occur in pairs, one set inherited from the mother and one set inherited from the father.
- Homologous chromosomes contain similar, but not identical, genetic information.
- Chromosome duplication is only one part of the cell cycle, which is a larger process consisting of growth, DNA replication, and cell division. When a cell is not dividing, it is in interphase.
- Actual cell division occurs during the mitotic phase of the cell cycle.

9.4 Mitosis and Cytokinesis
- Mitosis and cytokinesis occur as part of the sequence of events that define cell function—the cell cycle. The cell cycle consists of two parts—a long interphase, during which the cell grows and carries out its specific functions (including replicating its DNA), and a shorter mitotic phase, during which the duplicated DNA is evenly divided between two daughter cells created as the parental cell splits.
- The mitotic phase is subdivided into four parts—prophase, metaphase, anaphase, and telophase.
- Prophase and metaphase prepare the cell to divide its genetic material equally.
- Specialized microtubule structures form during prophase that define two cellular poles and help align the duplicated chromosomes (each composed of sister chromatids) along an equatorial plane during metaphase.
- Chromosomes are pulled apart during anaphase and are moved toward the opposite poles along the microtubules.

9.5 Variations in Cell Division
- Because plants have cell walls, the final phase of cell division (cytokinesis) is carried out differently; plants form a cell plate, which will later form the new cell wall.
- Prokaryotic cells such as bacteria have only a single circular chromosome, and they undergo a process called binary fission.

WORD ROOTS

bi- = two; **fission** = dividing or splitting (e.g., during *bi*nary *fission*, a bacterium splits into two cells)

chroma- = color; **soma-** = body (e.g., a *chromosome* is literally a colored body within the nucleus of the cell)

cyto- = cell or cells; **kinesis** = motion or movement (e.g., *cytokinesis* is the division of the cytoplasm of a cell after nuclear division)

homo- = the same (e.g., *homo*logous chromosomes are two chromosomes [a pair] that have corresponding genetic information)

karyo- = nucleus (e.g, *karyo*type is a photograph [print] of the number and size of the chromosomes found in the nucleus)

KEY TERMS

binary fission _____

_____

cell cycle _____

_____

centrosome _____

_____

chromatid _____

_____

chromatin _____

_____

chromosome _____

_____

cytokinesis _____

_____

enzyme _____

_____

genetics _____

_____

genome _____

_____

homologous chromosomes _____

_____

interphase _____

_____

karyotype _____

_____

metaphase plate _____

_____

microtubule _____

_____

mitosis _____

_____

mitotic phase _____

_____

mitotic spindle _____

_____

FLASH CARDS

To use the flash cards, tear the page from the book and cut along the dashed lines. The key term appears on one side of the flash card, and its definition appears on the opposite side.

binary fission	cytokinesis
cell cycle	enzyme
centrosome	genetics
chromatid	genome
chromatin	homologous chromosomes
chromosome	interphase

the physical separation of one cell into two daughter cells

the form of reproduction carried out by prokaryotic cells in which the chromosome replicates and the cell pinches between the attachment points of the two resulting chromosomes to form two new cells; in this type of simple cell splitting, each pair of daughter cells is an exact replica of the parental cell

a chemically active protein that speeds up, or in practical terms enables, chemical reactions in living things

the repeating pattern of growth, genetic duplication, and division seen in most cells

the study of physical inheritance among living things

a cellular structure that acts as an organizing center for the assembly of microtubules; a cell's centrosome duplicates prior to mitosis and plays an important part in the development of the cell's mitotic spindle

the complete collection of an organism's genetic information; more narrowly, the complete haploid set of an organism's chromosomes

one of the two identical strands of chromatin (DNA plus associated proteins) that make up a chromosome in its duplicated state

chromosomes that are the same in size and function; species that are diploid (have two sets of chromosomes) have matching pairs of homologous chromosomes: one member of each homologous pair is inherited from the male, and the second member of each homologous pair is inherited from the female

a molecular complex of DNA and its associated proteins that makes up the chromosomes of eukaryotic organisms

that portion of the cell cycle in which the cell simultaneously carries out its work and—in preparation for division—duplicates its chromosomes; the other primary phase of the cell cycle is the mitotic phase (or M phase), which includes both mitosis (in somatic cells) and cytokinesis

structural unit containing part or all of an organism's genome, consisting of DNA and its associated proteins (chromatin); the human genome is made up of 23 pairs of chromosomes, or 46 chromosomes in all

karyotype

mitosis

metaphase plate

mitotic phase

microtubule

mitotic spindle

the separation of a somatic cell's duplicated chromosomes prior to cytokinesis

a pictorial arrangement of a full set of an organism's chromosomes

(M phase) that portion of the cell cycle that includes both mitosis and cytokinesis: mitosis is the separation of a somatic cell's duplicated chromosomes; cytokinesis is the physical separation of one cell into two daughter cells

a plane located midway between the poles of a dividing cell

the microtubules active in cell division, including those that align and move the chromosomes

the largest of the cytoskeletal filaments, microtubules take the form of hollow tubes composed of the protein tubulin; they help give structure to the cell, serve as the "rails" on which transport vesicles move, and form the cellular extensions known as cilia and flagella

SELF TEST

Once you have finished studying this chapter, close your books, grab a pencil, and spend the next 15 to 20 minutes completing this practice test.

Compare and Contrast

For each of the following paired terms, write a sentence of comparison ("Both") and a sentence of contrast ("However,").

chromatin/chromosome
microtubule/mitotic spindle
cell cycle/interphase
chromatid/homologous chromosomes
binary fission/mitosis

Short Answer

1. A sixth-grader wants to know how a group of molecules such as found in DNA could possibly contain instructions to build a whole human body. How would you explain this?

2. Why don't cells simply get larger and larger instead of "spending" the energy involved in cell division?

3. Human DNA consists of about 3 billion nitrogen bases. Given that cells are so small, explain how this amount of DNA can be found in a cell.

4. Fill out the following table describing the key events both within the nucleus and in the cytoplasm of each stage of mitosis and cytokinesis.

Stage	Key Events
Prophase	
Metaphase	
Anaphase	
Telophase	
Cytokinesis	

5. Draw a plant cell at the end of cell division. Label the cell plate, and explain why it is different from animal cells.

Multiple Choice

Circle the letter that best answers the question.

1. Which of the following best represents the relationship between genes and the genome?
   a. cell walls and cell membranes
   b. cars and streets
   c. books and libraries
   d. molecules and compounds
   e. acids and bases

Match the following terms with their description. Each choice may be used once, more than once, or not at all.

   a. cytokinesis
   b. chromosome
   c. chromatin
   d. centrosome
   e. genome

2. _____ Combination of DNA and protein

3. _____ The complete collection of genetic information

4. _____ The splitting of one cell into two

5. _____ The spindle-organizing structure during cell division

6. _____ Individual packets of DNA

7. All of the following are DNA bases *except:*
   a. adenine.
   b. cytosine.
   c. guanine.
   d. alanine.
   e. thymine.

8. Which of the following is the last phase of mitosis?
   a. anaphase
   b. telophase
   c. prophase
   d. metaphase
   e. cytokinesis

9. If a cell has 30 chromosomes before mitosis, how many does each daughter cell have afterward?
   a. 10
   b. 15
   c. 20
   d. 25
   e. 30

10. The sequence of DNA specifies:
    a. nuclear structure.
    b. ribosome activity.
    c. amino acid sequence of proteins.
    d. a karyotype.
    e. what type of division the cell will undergo.

11. A human cell has _____ chromosomes.
    a. 22
    b. 23
    c. 44
    d. 46
    e. 48

12. Human males and females have different:
    a. sex chromosomes.
    b. numbers of chromosomes.
    c. karyotypes.

    d. somatosomes.

    e. autosomes.

13. At which stage of the cell cycle does DNA replication occur?

    a. G_2

    b. S

    c. G_1

    d. M

    e. cytokinesis

14. Plant cells divide by:

    a. cytokinesis without mitosis.

    b. mitosis without cytokinesis.

    c. binary fission.

    d. binary fusion.

    e. mitosis followed by cell-plate formation.

15. Which stage of division is associated with the action of the contractile ring?

    a. plant-cell cytokinesis

    b. animal-cell anaphase

    c. bacterial-cell cytokinesis

    d. animal-cell cytokinesis

    e. bacterial-cell binary fission

16. Why does mitosis precede cytokinesis?

    a. because prokaryotic cells require mitotic products to fuel cytokinesis

    b. because DNA takes up most of the inner cell space

    c. because it is more important to divide the outside of the cell before the inside

    d. because it is more important to divide the genetic material evenly before the cell infrastructure

    e. because most biological processes occur in reverse order

Read the following example and then answer questions 17–18.

You decide to give your memoirs to your two daughters. To ensure that each daughter receives an intact set, you copy each page and then file the original and the copy together, connected by a paper clip. When this task is finished, you systematically separate the files into two piles, thereby ensuring that each daughter receives one (and only one) copy of each page.

17. In this example, what is analogous to DNA?

    a. the daughters

    b. the paper clips

    c. the cells

    d. the memoirs

    e. the entire process

18. In this example, what is analogous to the mitotic stage known as anaphase?

    a. deciding to give the memoirs to your daughters

    b. copying each page

    c. connecting the pages with paper clips

    d. systematically separating the files into two piles

    e. the two files at the end of the process

19. In telophase of mitosis, the mitotic spindle dissolves. This is essential because it is the opposite of what occurred in:

    a. prophase.

    b. interphase.

    c. S phase.

    d. anaphase.

    e. M phase.

20. The mitotic spindle is made of:

    a. Golgi bodies.

    b. microtubules.

    c. mitochondria.

    d. chromosomes.

    e. DNA.

21. If cytokinesis does not occur during cell division, the resulting cells would probably:

    a. have unequal amounts of cytoplasm.

    b. not have the correct number of chromosomes.

    c. not have the correct amount of chromatin.

    d. be very large.

    e. be cancerous.

22. A human nerve cell, in prophase of miotis, contains 46 chromosomes. How many chromatids does it contain altogether?

    a. 46

    b. 92

    c. 23

    d. 69

    e. 23 or 46, depending on when during prophase you look at the cell

23. The number of chromosomes found in a eukaryotic cell:

    a. can vary if the organism is young or old.

    b. is constant during the entire life.

    c. can vary from cell to cell.

    d. will be larger if the organism is an animal, rather than a plant.

    e. will increase during mitosis.

24. Cancer cells are characterized by an accelerated rate of cell division. All of the following are ways that researchers are studying cancer cells *except:*

    a. investigating the mechanisms by which the cell cycle becomes acclerated.

    b. looking at the proteins that control the cell cycle.

    c. looking at the genetic makeup of cancer cells.

    d. looking for DNA damage.

    e. using chemicals to mutate cancer cells.

25. A karyotype:

    a. compares one set of chromosomes to another.

    b. is a picture of chromosomes arranged according to size.

    c. of a normal human cell shows 48 chromosomes.

    d. of a normal female will have both an X and a Y chromosome.

    e. can be used to look at the genes on a particular chromosome.

WHAT'S IT ALL ABOUT?

Beginning with this unit, we are going to start providing you with fewer aids to guide you in answering these questions. The answer key will still provide direction about what kind of question is being asked and the type of evidence appropriate to answer the question, but even that will start to become less detailed. Why? Because we are confident that after doing these questions for the first eight chapters, you don't need any more than a nudge in the right direction.

Question: The key experiments revealing that DNA indeed contains the "recipes" for all cellular components took place only 50 to 60 years ago. Until those classic experiments, many scientists thought that proteins might be the carrier of genetic information. What would be the advantages and disadvantages of proteins as information storage molecules?

1. **What type of question is this?**
   Compare and contrast? Defend a position? Describe the effect of "A" on "B"?

2. **What kind of evidence do you need?**
   This depends in part on what type of question you think it is. However, given that DNA and protein both appear in the question, maybe you need to think about the structures of these molecules and how structure affects function.

3. **Pull it all together.**
   The question asks for "advantages and disadvantages"—that request should shape your answer.

CHAPTER 10 PREPARING FOR SEXUAL REPRODUCTION: MEIOSIS

Basic Chapter Concepts

* In meiosis, chromosome duplication is followed by two division steps that produce four haploid cells, each with a single copy of each chromosome.
* Meiosis ensures genetic diversity through recombination of homologous chromosome pairs and the independent assortment of chromosomes during meiosis I.
* Meiosis reduces the number of chromosomes by half, making sexual reproduction possible, although not all organisms reproduce sexually.

CHAPTER SUMMARY

10.1 An Overview of Meiosis
* Meiosis reduces a diploid parental cell into four haploid daughter cells because the single round of DNA duplication is followed by two rounds of cell division, each of which divides the DNA equally among the daughter cells.

10.2 The Steps in Meiosis
* The steps of meiosis I involve matching and separation—chromosome pairs are aligned along the metaphase plate, and then one of each chromosome pair is randomly assorted into each of the two daughter cells.
* Meiosis II consists of cell division without DNA duplication. The sister chromatids of each chromosome are pulled apart during anaphase II into the daughter cells (similar to division during mitosis).

10.3 What is the Significance of Meiosis?
* Meiosis produces gametes, haploid cells that can fuse to make a diploid zygote, thus avoiding the problem of genome duplication each time sexual reproduction occurs.
* Meiosis increases genetic diversity. Chromosome pairs can intertwine and exchange pieces when pairing during prophase I. Also, because chromosome pairs align along the equatorial plane independently during anaphase I, maternally and paternally derived chromosomes are mixed within the daughter cells.

10.4 Meiosis and Sex Outcome
* In humans, the X and Y chromosomes segregate during meiosis, producing gametes that bear either an X chromosome or a Y chromosome. Sex is determined when the zygote forms—fusion of two X-bearing gametes produces the human female, and fusion of one X-bearing gamete and one Y-bearing gamete produces the human male. Fusion of two Y-bearing gametes is not possible.

10.5 Gamete Formation in Humans
* Males produce sperm in the testes from precursor cells (spermatogonia). Sperm are an efficient delivery system for 22 autosomes and either an X or Y chromosome.
* Females produce eggs from precursor cells (oogonia) with sufficient cellular content to support the growth of the zygote produced following fusion with a sperm. Each egg contains 22 autosomes and an X chromosome.

10.6 Life Cycles: Humans and Other Organisms
* All sexually reproducing organisms exist in two sexes and produce gametes that fuse to make a new organism.
* Some organisms reproduce asexually, producing offspring from the parent through a mitotic process—binary fission in bacteria and vegetative reproduction in plants.

WORD ROOTS

-ploid = number (e.g., di*ploid* cells have nuclei with two copies of each chromosome)

-cyte = cell (e.g., an adipo*cyte* is a fat-storage cell)

oo- = egg (the *oo*cyte is the egg cell)

KEY TERMS

n _____

2n _____

asexual reproduction _____

crossing over _____

diploid _____

gamete _____

haploid _____

independent assortment _____

life cycle _____

meiosis _____

oogonia _____

polar body _____

primary oocyte _____

primary spermatocyte _____

sex chromosome _____

sexual reproduction _____

spermatogonia _____

tetrad _____

FLASH CARDS

To use the flash cards, tear the page from the book and cut along the dashed lines. The key term appears on one side of the flash card, and its definition appears on the opposite side.

n	meiosis
2n	oogonia
asexual reproduction	polar body
crossing over	primary oocyte
diploid	primary spermatocyte
gamete	sex chromosome
haploid	sexual reproduction
independent assortment	spermatogonia
life cycle	tetrad

a process in which a single diploid cell divides to produce four haploid reproductive cells

in a living cell, the condition of being haploid: of having one set of chromosomes

the diploid cells that are the starting female cells in gamete (egg) production; diploid oogonia develop into diploid primary oocytes, which give rise to haploid secondary oocytes

in a living cell, the condition of being diploid: of having two sets of chromosomes

nonfunctional cell produced during meiosis in females

reproduction that occurs without the union of two reproductive cells (sexual reproduction); offspring produced through asexual reproduction are genetically identical to their parent organism

a diploid cell produced in females that may mature into an egg, initially by giving rise to haploid secondary oocytes; after the female reaches puberty, an average of one oocyte per month is selected to continue the process of maturation in the ovary

(genetic recombination) a process, occurring during meiosis, in which homologous chromosomes exchange reciprocal portions of themselves

a diploid cell in a male that will undergo meiosis to produce haploid secondary spermatocytes, which ultimately give rise to mature sperm cells

possessing two sets of chromosomes: all human cells are diploid, with the exception of human gametes (eggs and sperm), which are haploid; such haploid cells possess only a single set of chromosomes

the chromosomes that determine the sex of an organism; the X or Y chromosomes in humans

a haploid reproductive cell, either egg or sperm

a means of reproduction in which the nuclei of the reproductive cells from two separate organisms fuse to produce offspring

possessing a single set of chromosomes; human gametes (eggs and sperm) are haploid cells because they have only a single set of chromosomes; all other cells in the human body are diploid, meaning they possess two sets of chromosomes

diploid cells that are the starting cells in sperm production in males; spermatogonia are reproductive stem cells in that, in dividing, each of them produces one primary spermatocyte (which will develop into four mature sperm cells) and one spermatogonium

the random distribution of homologous chromosome pairs on differing sides of the metaphase plate during meiosis

the grouping formed by the linkage of two homologous chromosomes in prophase I of meiosis; the four sister chromatids involved in this linkage give it the name *tetrad*

the repeating series of steps that occurs in the reproduction of an organism

SELF TEST

Once you've finished studying this chapter, close your books, grab a pencil, and spend 15 to 20 minutes working on this practice test.

Compare and Contrast

For each of the following paired terms, write a sentence of comparison ("Both") and a sentence of contrast ("However,").

chromosome/chromatid
haploid/diploid
meiosis/mitosis
spermatocytes/oogonia
meiosis I/meiosis II

Short Answer

1. How common is meiosis when compared to mitosis? Why?

2. Why is meiosis II necessary?

3. What is the significance of meiosis?

4. What determines the sex of offspring in humans?

5. In humans, males can produce many more sperm than females can produce eggs. Explain.

6. What are the advantages for organisms that reproduce asexually?

Multiple Choice

Circle the letter that best answers the question.

1. Which of the following is true regarding sperm?
   a. They have all the typical organelles.
   b. One is produced for every spermatogonium.
   c. They are haploid.
   d. They are much larger than human eggs.
   e. They are somatic cells.

2. The sex of a human fetus is determined by:
   a. the age of the mother.
   b. the age of the father.
   c. the phase of the moon at the time of conception.
   d. the sex chromosome found in the sperm.
   e. the sex chromosome found in the egg.

3. In humans, meiosis produces cells that are:
   a. identical.
   b. haploid.
   c. gametes.
   d. b and c
   e. all of the above

4. The process of growing a plant cutting into a plant is an example of:
   a. asexual reproduction.
   b. sexual reproduction.
   c. meiosis.

d. oogenesis.
e. spermatogenesis.

5. During meiosis, homologous chromosomes separate at:
   a. prophase I.
   b. metaphase I.
   c. anaphase I.
   d. metaphase II.
   e. anaphase II.

6. Which of the following allows sexual reproduction to occur generation after generation?
   a. mitosis
   b. mucosis
   c. meiosis
   d. cytosis
   e. halitosis

7. Recombination occurs:
   a. during mitosis I.
   b. during anaphase II.
   c. during prophase I.
   d. during fertilization.
   e. during binary fission.

8. Why are no two gametes exactly alike?
   a. because their cell membranes are unique, like snowflakes
   b. because each gamete undergoes a slightly different version of mitosis during its formation
   c. because each gamete has a different combination of parental chromosomes
   d. because each gamete is the result of crossing over
   e. c and d are correct

9. In meiosis II:
   a. chromatids are separated and sent to separate daughter cells.
   b. homologous chromosomes are separated into different daughter cells.
   c. haploid chromosomes fuse to make a diploid cell.
   d. bacteria make exact copies of their chromosomes.
   e. crossing over occurs.

10. Meiosis is:
    a. shared by all living organisms.
    b. found only among bacteria.
    c. found only in sexually reproducing organisms.
    d. found in every cell of the body.
    e. found only among mammals.

11. Reginald Prettiboy decides to clone himself. What sex would the "offspring" be?
    a. male
    b. female
    c. hermaphrodite
    d. either male or female
    e. asexual

12. Which of the following is *not* common to both oogenesis and spermatogenesis?
    a. They are both formed by two rounds of meiosis.
    b. They both begin with a diploid cell.
    c. They both produce gametes.
    d. They both result in the formation of four haploid gametes.
    e. They both occur in mammals.

13. You discover a new type of animal that you have named *Quinntella georgiana*. Georgiana is a type of sea star, which means that it is capable of reproduction by:
    a. regeneration.
    b. hermaphroditism.
    c. binary fission.
    d. vegetative reproduction.
    e. meiosis.

14. Which of the following is true about recombination?
    a. It resembles mitosis.
    b. It involves sperm production.
    c. It requires that homologous chromosomes pair up.
    d. It produces haploid structures that do not complete gamete formation.
    e. It is a process by which genetic material is exchanged between homologous chromosomes.

15. Which of the following is true about polar body formation?
    a. It resembles mitosis.
    b. It involves sperm production.
    c. It requires that homologous chromosomes pair up.
    d. It produces haploid structures that do not complete gamete formation.
    e. It is a process by which genetic material is exchanged between homologous chromosomes.

16. Which of the following is true about spermatogenesis?
    a. It resembles mitosis.
    b. It involves sperm production.
    c. It does not require that homologous chromosomes pair up.
    d. It produces haploid structures that do not complete gamete formation.
    e. It is a process by which genetic material is exchanged between homologous chromosomes.

17. Which of the following is true about meiosis II?
    a. It resembles mitosis.
    b. It involves sperm production.
    c. It requires that homologous chromosomes pair up.
    d. It produces haploid structures that do not complete gamete formation.
    e. It is a process by which genetic material is exchanged between homologous chromosomes.

18. Which of the following is true about prophase I?
    a. It resembles mitosis.
    b. It involves sperm production.
    c. It requires that homologous chromosomes pair up.
    d. It produces haploid structures that do not complete gamete formation.
    e. It is a process by which genetic material is exchanged between homologous chromosomes.

19. How does meiosis generate genetic diversity?
    a. Homologous chromosomes cross over during prophase I.
    b. During metaphase I, chromosomes align randomly.
    c. During prophase II, chromosomes exchange material.
    d. a and b
    e. b and c

20. If a cell begins meiosis with a 2n (diploid) number of 16, how many chromosomes would there be in prophase I?
    a. 8
    b. 16
    c. 32
    d. 64
    e. 128

21. If a cell begins meiosis with a 2n (diploid) number of 16, how many chromosomes would there be in anaphase I?
    a. 8
    b. 16
    c. 32
    d. 64
    e. 128

22. If a cell begins meiosis with a 2n (diploid) number of 16, how many chromosomes would there be in metaphase II?
    a. 8
    b. 16
    c. 32
    d. 64
    e. 128

23. If a cell begins meiosis with a 2n (diploid) number of 16, how many chromosomes would there be in anaphase II?
    a. 8
    b. 16

c. 32

d. 64

e. 128

24. If a cell begins meiosis with a 2n (diploid) number of 16, how many chromosomes would there be in telophase II?

a. 8

b. 16

c. 32

d. 64

e. 128

25. Diploid cells become haploid during:

a. meiosis I.

b. meiosis II.

c. mitosis.

d. a and c

e. b and c

WHAT'S IT ALL ABOUT?

Here's a question to help you pull together what you've learned so far using this text.

Question: Meiosis eliminates one problem of sexual reproduction by reducing by half the number of chromosomes in the gametes. Sometimes, however, this dance of the chromosomes goes awry when the chromosomes are not divided evenly among the four resulting gametes. In Chapter 12, we will consider in more detail what happens when gametes end up with too many or too few chromosomes, but for now let's think about how this error might happen. Think about how the process of meiosis proceeds when making human sperm and oocytes; when or how do you think it is most likely that chromosomes will not segregate properly? To help you with your thinking, consider the observation that missegregation of chromosomes happens more often in older humans.

1. **What type of question is this?**

   Compare and contrast? Defend a position? Describe the effect of "A" on "B"?

2. **What kind of evidence do you need?**

   Consider the key parts of this question: chromosomal missegregation in meiosis, and the process of meiosis in human gamete formation. This question appears to be limited to the information you found in this chapter.

3. **Pull it all together.**

   You can do it! This question is probably a lot harder than any you would find on an exam at this stage of the course, so it is really good practice for your critical-thinking skills.

CHAPTER 11 THE FIRST GENETICIST: MENDEL AND HIS DISCOVERIES

Basic Chapter Concepts

- The physical appearance of an organism, the phenotype, is determined by the function and interactions of the proteins encoded by the organism's genome (genotype).
- Gregor Mendel's quantitative studies on the inheritance of physical traits in pea plants first revealed three basic laws of genetics:
  - Genetic elements come in pairs.
  - Genetic elements separate when gametes are made.
  - Genetic elements segregate independently of each other.
- Mendel's laws can still account for variations in phenotype caused by protein interactions and environmental effects.

CHAPTER SUMMARY

11.1 Mendel and the Black Box
- Mendel's achievement was to infer principles that explain the heritability of traits by matching inputs with outputs, without knowing anything about the physical mechanism involved in genetic information transfer.
- *Phenotype* refers to the visible characteristics of an organism, its actions, function, size, shape, and so on. The proteins encoded by an organism's genes determine these observable traits.

11.2 The Experimental Subjects: *Pisum sativum*
- Mendel's experimental organism, *Pisum sativum*, has seven discernible traits; seed color and shape, pod color and shape, flower color and location, and plant height. Plants can be bred true for each trait, so the "inputs" can be determined with absolute certainty.

11.3 Starting the Experiments: Yellow and Green Peas
- By counting how often a trait appears among the offspring of a cross of true-breeding parental plants, Mendel quantified the frequency of each trait within the filial generation.
- From these proportions, Mendel inferred that:
  - Traits come in pairs, and one trait of the pair "dominates" over the other.
  - Traits do not "blend" (show an intermediate phenotype), because the "recessive" trait can always be recovered when plants of the first filial (F_1) generation are crossed.

11.4 Another Generation for Mendel
- After Mendel found that recessive phenotypes not evident in the F_1 generation reappeared in the F_2 generation, he allowed the seeds of the F_2 generation to self-pollinate to produce the F_3 generation.
- Results from the F_3 generation showed only two phenotypes, which actually represented three different genotypes. Mendel reasoned that differing characters in organisms result from two genetic elements (alleles) that separate when gametes are formed during meiosis. This is known as Mendel's First Law of Law of Segregation.
- Organisms that have two identical copies of an allele are homozygous for the trait. An organism with two different alleles is heterozygous for the trait.

11.5 Crosses Involving Two Characters
- In a dihybrid cross involving two heterozygous characters (traits), Mendel observed a specific distribution of genotypes (9:3:3:1 ratio). From this ratio, Mendel reasoned that the characters were transmitted independent of one another.

- The basis for Mendel's Second Law, the Law of Independent Assortment, is that during meiosis the genetic elements responsible for each trait, which we now refer to as the gene responsible for each phenotype, segregate into the gametes independently of each other.

11.6 Reception of Mendel's Ideas
- Mendel presented his work to a local scientific society in 1865, but his work was poorly received because the scientific community at the time did not grasp the significance of his investigations.
- Mendel's work was rediscovered in 1900 and became the basis for the field of genetics.

11.7 Incomplete Dominance
- Incomplete dominance of one trait over another occurs when one allele of the gene responsible for the phenotype is nonfunctional—it does not produce a product, as in the snapdragon example in which one "white flower" allele does not produce a pigment protein. The phenotype of the heterozygous organism seems to be intermediate between the parental types, but it simply reflects the presence of half of the concentration of functional protein.

11.8 Lesson from Blood Types: Codominance
- Codominant traits result when different alleles of a single gene each produce functional proteins. As in the A and B blood types, when both alleles are present, both proteins are made, and neither protein "dominates" over the other.

11.9 Multiple Alleles and Polygenic Inheritance
- It is rare for a single gene to produce a single phenotype; most gene products interact with each other, producing multiple phenotypes.

11.10 Genes and Environment
- Gene expression can be affected by environmental conditions, producing variable phenotypes among genetically identical organisms.

11.11 One Gene, Several Effects: Pleiotropy
- A single gene can have several wide-reaching effects. For example, in fragile-X syndrome, a defect in one gene results in a wide spectrum of physical and mental abnormalities.

WORD ROOTS

di- = two (e.g, a *di*hybrid cross is a cross involving two different traits)

geno- = offspring (e.g., *geno*type is the genetic makeup of an individual)

hetero- = different (e.g., *hetero*zygous refers to having nonidentical alleles for a specific gene)

homo- = same (e.g, *homo*zygous refers to having identical alleles for a specific gene)

pheno- = to show (e.g., *pheno*type is the physical appearance of a trait)

poly- = many (e.g., *poly*genic inheritance is the inheritance of multiple genes that affect the same trait, such as height in humans)

KEY TERMS

allele _____

_____

bell curve _____

_____

codominance _____

_____

cross-pollinate _____

_____

dihybrid cross _____

_____

dominant _____

_____

first filial generation (F$_1$) _____

_____

genotype _____

_____

heterozygous _____

_____

homozygous _____

_____

incomplete dominance _____

_____

Law of Independent Assortment _____

_____

Law of Segregation _____

_____

monohybrid cross _____

_____

multiple alleles _____

_____

parental generation (P) _____

_____

phenotype _____

_____

pleiotropy _____

_____

polygenic inheritance _____

_____

recessive _____

_____

rule of addition _____

_____

rule of multiplication _____

_____

FLASH CARDS

To use the flash cards, tear the page from the book and cut along the dashed lines. The key term appears on one side of the flash card, and its definition appears on the opposite side.

allele	first filial generation (F$_1$)
bell curve	genotype
codominance	heterozygous
cross-pollinate	homozygous
dihybrid cross	incomplete dominance
dominant	Law of Independent Assortment

the offspring of the parental generation in an experimental cross

---

one of the alternative forms of a single gene; in pea plants, a single gene codes for seed color, and it comes in two alleles—one codes for yellow seeds, the other for green seeds

---

the genetic makeup of an organism, including all the genes that lie along its chromosomes

---

a distribution of values that is symmetrically largest around the average

---

possessing two different alleles of a gene for a given character

---

a condition in which two alleles of a given gene have different phenotypic effects, with both effects manifesting in organisms that are heterozygous for the gene

---

having two identical alleles of a gene for a given character

---

to pollinate one plant with pollen of another plant; Mendel used this technique in conducting his experiments to uncover rules of heredity

---

a genetic condition in which the heterozygote phenotype is intermediate between either of the homozygous phenotypes

---

an experimental cross in which the plants differ in two of their characters

---

during gamete formation, gene pairs assort independently of one another; also known as Mendel's Second Law, this is one of the principles of inheritance formulated by Gregor Mendel

---

term used to designate an allele that is expressed in the heterozygous condition

Law of Segregation

pleiotropy

monohybrid cross

polygenic inheritance

multiple alleles

recessive

parental generation (P)

rule of addition

phenotype

rule of multiplication

the phenomenon by which one gene has many effects

---

differing characters in organisms result from two genetic elements (alleles) that separate in gamete formation, such that each gamete gets only one of the two alleles; also known as Mendel's First Law, this is one of the principles of inheritance formulated by Gregor Mendel

---

inheritance of a genetic character that is determined by the interaction of multiple genes, with each gene having a small additive effect on the character

---

an experimental cross in which organisms are tested for differences in one character

---

term used to designate an allele that is not expressed in the heterozygous condition

---

three or more alleles—alternative forms of a gene—occurring in a population

---

in probability theory, the principle that when an outcome can occur in two or more different ways, the probability of that outcome is the sum of the respective probabilities

---

the generation that begins an experimental cross between organisms; such a cross is used to study genetics and heredity of traits

---

in probability theory, the principle that the probability of any two events happening is the product of their respective probabilities

---

a physical function, bodily characteristic, or action of an organism

SELF TEST

Once you have finished studying this chapter, close your books, grab a pencil, and spend the next 15 to 20 minutes completing this practice test.

Compare and Contrast

For each of the following paired terms, write a sentence of comparison ("Both") and a sentence of contrast ("However").

genotype/phenotype
allele/gene
Law of Segregation/Law of Independent Assortment
dominant/recessive
rule of addition/rule of multiplication

Short Answer

1. Why was Mendel able to figure out the principles of heredity while many other investigators failed to do so?

2. A mutation in a chloride channel gene is responsible for cystic fibrosis, a recessive genetic disorder in which the lungs are filled with a thick mucus that impairs their function. How would you describe the genotype and phenotype of a person with cystic fibrosis?

3. You find a new plant species, some having orange flowers and some having yellow flowers. When you cross a yellow- and an orange-flowered plant, all the offspring have orange flowers. Which flower color is dominant? If flower color behaves the same way as pea color, what ratio of yellow- and orange-flowered plants do you expect in the F_2 generation?

4. Two parents, both of whom are heterozygous for the cystic fibrosis gene mutation, have four children. The inheritance of cystic fibrosis follows Mendel's laws.
   a. Draw the Punnett square for these parents.
   b. On average, how many children (out of four) would you expect to be homozygous for the cystic fibrosis gene mutation?

5. Individuals with Marfan syndrome experience several health problems, including cardiac, eye, and joint problems. What is an explanation for this?

6. Not every woman with the "breast cancer" gene mutation will develop breast cancer. Why might this be?

7. Why are organisms such as peas and fruit flies better subjects for genetics investigations than human beings?

8. The inheritance of flower color in snapdragons shows incompelote dominance: when a red snapdragon is crossed with a white one, all their offspring are pink. What offspring would be produced, in what proportions, if two of these pink snapdragons were crossed? Use a Punnett square to show your results.

9. If you flipped two coins, the probabilty that you will get two tails is 1:4. However, the probability of getting one head and one tail is 1:2. Explain why.

10. Identical twins were adopted by different parents at birth. The twins were reunited as adults, and when their IQ scores were compared, one had an IQ of 125 and the other an IQ of 110. Given that they both have the same genes, how could this difference be accounted for?

11. If you were diagnosed with a genetic disease, what specific questions might you ask your physician, given what you have learned from this chapter?

Multiple Choice

Match the following terms with their description. Each choice may be used once, more than once, or not at all.
   a. alleles
   b. B genes
   c. phenotype
   d. pleiotropy
   e. codominance

1. _____ Physical appearance of an organism

2. _____ When alleles cannot mask or "cover" each other

3. _____ Different versions of the same gene

4. _____ Specify instructions for building proteins

5. _____ When one gene affects multiple aspects of the phenotype

Circle the letter that best answers the question.

6. You perform a cross between red ladybugs and white ladybugs and see all pink ladybugs. What is the relationship between the red and white ladybug color?
   a. Red is dominant to white.
   b. Red is incompletely dominant.
   c. White is dominant to red.
   d. Red is pleiotropic.
   e. none of the above

7. You carry out a self cross of the F_1 of a yellow and green pea monohybrid cross. Which of the following phenotypic ratios do you expect to see in the F_2?
   a. 9:3:1
   b. 1:2:1
   c. 3:1
   d. 9:3:3:1
   e. 9:3:4

8. What do the terms *gene* and *allele* have in common?
   a. Both refer to the type of pea plant Mendel developed.
   b. Both could refer to the same DNA sequence.
   c. Both refer to Greek words.
   d. Both refer to the blending of inherited traits.
   e. They have nothing in common.

9. A single allele of a gene that exerts multiple phenotypic effects is due to:
   a. multiple alleles.
   b. interactions with the environment.
   c. pleiotropy.
   d. independent assortment.
   e. segregation.

10. In a monohybrid cross, a 3:1 phenotypic ratio is observed in the F_1 generation. What is the underlying genotypic ratio?
    a. 3:1
    b. 9:3:3:1
    c. 1:1
    d. 1:2:1
    e. 1:2

11. To have cystic fibrosis, a person must be homozygous recessive for the disease allele. If two heterozygous parents have many children, what is the expected ratio of children with cystic fibrosis to those without cystic fibrosis?
    a. three with and one without
    b. one with and three without
    c. four with and one without
    d. one with and four without
    e. all of them will have it

12. The major difference between dominant and recessive alleles of the same trait is that:
    a. when both are present, only the recessive alleles are expressed.
    b. when both are present, neither is expressed.
    c. when both are present, only the dominant is expressed.
    d. when neither is present, both are expressed.
    e. dominant traits are more useful than recessive traits.

13. Which of the following is not a possible offspring type from the mating of AA and Aa?
    a. AA
    b. Aa

    c. aa
    d. all of the above
    e. none of the above

14. When Mendel followed the inheritance of two traits at the same time, he found that:
    a. the inheritance pattern of one trait would always control the other.
    b. no conclusions could be drawn from his results.
    c. the traits passed from one generation to the next independently of each other.
    d. one trait increased the expression of the other.
    e. it was impossible to sort out the offspring of the crosses.

15. Reginald Pisum is homozygous dominant for the trait Sativum, or pea-like earlobes. If he were to marry Pearl Blossom, who is heterozygous for the trait, what proportion of their offspring will also be homozygous dominant?
    a. 0
    b. 0.25
    c. 0.50
    d. 0.75
    e. 1.00

16. Reginald Pisum is homozygous dominant for the trait Sativum. How did he inherit his alleles?
    a. both dominants from his mother
    b. both dominants from his father
    c. dominant from his mother, dominant from his father
    d. dominant from his mother, recessive from his father
    e. dominant from his father, recessive from his mother

17. What is the phenotypic ratio for the offspring of a cross between a homozygous dominant (ZZ) and a homozygous recessive (zz)?
    a. all dominant
    b. all recessive
    c. half dominant, half recessive
    d. one dominant, one heterozygous, two recessive
    e. one dominant, two heterozygous, one recessive

18. How did Mendel's studies in genetics differ from earlier studies of breeding and inheritance?
    a. Mendel worked with plants, whereas earlier studies dealt with animals.
    b. Mendel's work was more quantitative.
    c. Mendel worked with wild species, not domesticated species.
    d. Mendel picked traits that were on different chromosomes.
    e. Mendel found that offspring differ significantly from their parents.

19. Which of the following is *not* a characteristic of the pea plants with which Mendel worked?
    a. They produced male and female parts.
    b. They exhibited a blending of characteristics.
    c. They normally would self-fertilize.
    d. They had many different recognizable traits.
    e. They were easy to cultivate.

20. In a Punnett square, the letters outside the little boxes represent:
    a. gametes.
    b. offspring genotypes.
    c. offspring phenotypes.
    d. parental phenotypes.
    e. parental genotypes.

21. The Law of Segregation:
    a. deals with the alleles governing two different traits.
    b. applies only to genes on the same chromosome.
    c. indicates that the expression of one gene is independent of the expression of another gene.
    d. refers to the fact that alleles for the same characteristic separate during meiosis.
    e. was true only for Mendel's experiments.

22. According to Mendel's First Law, the gametes of a heterozygous individual will be:
    a. all dominant alleles.
    b. all recessive alleles.
    c. 50 percent dominant and 50 percent recessive.
    d. 25 percent dominant alleles.
    e. 25 percent recessive alleles.

23. A new family moved into your neighborhood. They have four children, all of whom are boys, and the mother is pregnant. What is the probability that the new baby will be a boy?
    a. less than 50 percent
    b. more than 50 percent
    c. exactly 50 percent
    d. Because there are already four boys, the new baby is most certainly a girl.
    e. It depends entirely on which month the birth occurs.

24. Figure 11.13 in your text shows a picture of a group of students, with shorter individuals on one side and taller individuals on the other. The fact that most students show up in the middle of the curve indicates:
    a. pleiotropy.
    b. Mendel's Law of Independent Assortment.
    c. Mendel's Law of Segregation.
    d. continuous variation of a trait.
    e. height is most likely a one-gene trait.

25. Mendel's Law of Independent Assortment was nearly impossible for most scientists to understand until there was a better understanding of:
    a. dominant traits.
    b. recessive traits.
    c. meiosis.
    d. mitosis.
    e. multiple alleles.

WHAT'S IT ALL ABOUT?

Here's a question to help you pull together what you've learned so far using this text.

Question: The hallmark of Mendel's work was his careful record keeping and large sample size; his quantitative results led to identifying the phenotypic ratios that he used to infer the laws of independent assortment and segregation. Mendel was lucky, too, to have picked seven phenotypes that were all independent. What would have happened to his analysis if two of the seven traits were not independent—say, if all yellow seeds were always smooth, and all green seeds were always round?

1. **What type of question is this?**
   Compare and contrast? Defend a position? Describe the effect of "A" on "B"?

2. **What kind of evidence do you need?**
   What did Mendel's ratios tell him about the inheritance of traits? If five of the traits produced predictable ratios leading to the description of "dominant" and "recessive" traits but two did not, how do you think Mendel would have interpreted the results, given his knowledge base?

3. **Pull it all together.**
   Go for it!

CHAPTER 12 UNITS OF HEREDITY: CHROMOSOMES AND INHERITANCE

Basic Chapter Concepts

- Human sex chromosomes play a unique role in heredity.
- Malfunctioning chromosomes change the phenotype of the organism, a fact indicating that chromosomal makeup and function are critical to human health.
- Phenotype changes may be caused either by defective alleles, by physical damage to the chromosome, or by incorrect separation of chromosomes during gametogenesis.

CHAPTER SUMMARY

12.1 X-Linked Inheritance in Humans

- Defects in alleles on the X chromosome are expressed more often in males than in females because males have a single X chromosome; consequently, color blindness and hemophilia (diseases caused by defective X-chromosome genes) occur more often in men than in women.
- An "extra" or a missing sex chromosome causes pleiotropic effects, disrupting development of the embryo.

12.2 Autosomal Genetic Disorders

- Diseases caused by nonfunctional alleles may be recessive, requiring the presence of two defective alleles to produce the disease phenotype, or they may be dominant, producing the disease phenotype even when one functional allele is present.

12.3 Tacking Traits with Pedigrees

- Changes in gene sequences carried on the autosomes can also produce nonfunctional or poorly functioning proteins, causing human disease. These autosomal genetic diseases can be detected by examining family pedigrees.

12.4 Aberrations in Chromosomal Sets: Polyploidy

- Functional human cells are diploid with 23 pairs of chromosomes. Our haploid gametes contain only 23 individual chromosomes. Some organisms may have multiple sets of chromosomes (triploid, tetraploid, etc.), a condition known as polyploidy. Whereas plants may be polyploid, this condition is fatal in humans.

12.5 Incorrect Chromosome Number: Aneuploidy

- Disease can be caused by having more or fewer chromosomes than a normal diploid number, as in Down syndrome. This aneuploidy occurs when the chromosomes segregate improperly during meiosis.
- Mitosis can cause aneuploidy as well. In fact, these nondisjunction events may be the source of some cancers.

12.6 Structural Abberations in Chromosomes

- Deletions occur when pieces of chromosomes are lost during meiotic recombination. Large-scale deletions, such as loss of parts of chromosomes, are easily detected in karyotype spreads.
- Inversions and translocations, in which part of a chromosome is "flipped" or exchanged with part of another chromosome, respectively, are also detectable in karyotype spreads.
- Duplications occur when meiotic crossing over between homologous chromosomes is unequal. The result is that one chromosome of the pair loses sequences, while the other gains a duplicate sequence.

WORD ROOTS

-ploid = units or number (e.g., a di*ploid* cell contains two of each chromosome type)

-some = body or structure (e.g., a lyso*some* is a structure containing lysing enzymes, while an autosome is a non-sex chromosome)

KEY TERMS

aneuploidy _____

autosomal dominant disorder _____

autosomal recessive disorder _____

carrier _____

deletion _____

dominant disorder _____

Down syndrome _____

inversion _____

nondisjunction _____

pedigree _____

polyploidy _____

recessive disorder _____

translocation _____

FLASH CARDS

To use the flash cards, tear the page from the book and cut along the dashed lines. The key term appears on one side of the flash card, and its definition appears on the opposite side.

aneuploidy	inversion
autosomal dominant disorder	nondisjunction
autosomal recessive disorder	pedigree
carrier	polyploidy
deletion	recessive disorder
dominant disorder	translocation
Down syndrome	

a chromosomal abnormality that comes about when a chromosomal fragment that rejoins a chromosome does so with an inverted orientation

a condition in which an individual organism has either more or fewer chromosomes than is normally found in its species' full set; Down syndrome is the result of aneuploidy—generally three copies of chromosome 21, rather than the standard two

the failure of homologous chromosomes or sister chromatids to separate during meiosis, resulting in unequal numbers of chromosomes in the daughter cells; nondisjunction results in aneuploidy

a genetic disorder caused by a single faulty allele located on an autosomal (non-sex chromosome); Huntington disease is one example

a familial history of genetically transmissible conditions; generally takes the form of a diagram

a recessive dysfunction caused by a faulty allele on an autosome (non-sex chromosome); sickle-cell anemia is one example

a form of sympatric speciation in which one or more sets of chromosomes are added to the genome of an organism; human beings cannot survive in a polyploid state, but many plants flourish in it; polyploidy is a means by which speciation can occur (most often in plants) in a single generation

a person who does not suffer from a recessive genetic debilitation, but who carries an allele for the condition that can be passed along to offspring

a medical condition that will not occur when an organism possesses a single functional allele for a given trait; red-green color blindness is an example of a recessive disorder in that only one set of functional alleles need be present for normal color vision

a chromosomal condition in which a piece of a chromosome has been lost; occurs when a chromosomal fragment that breaks off does not rejoin any chromosome

the swapping of fragments by non-homologous chromosomes, resulting in gene sequences that are out of order on both chromosomes

genetic conditions in which a single faulty allele can cause damage, even when a second, functional allele exists

a disorder in humans in which affected individuals usually have three copies of chromosome 21 rather than the standard two; individuals with this syndrome have short stature, shortened life span, and low IQ

SELF TEST

After you have finished studying this chapter, close your books, grab a pencil, and spend the next 15 to 20 minutes working on this practice test.

Compare and Contrast

For each of the following paired terms, write a sentence of comparison ("Both") and a sentence of contrast ("However,").

aneuploidy/polyploidy
inversion/translocation
Turner/Klinefelter
sex-linked recessive/autosomal recessive
autosome/sex chromosome

Short Answer

1. What is different about the inheritance of traits located on the sex chromosomes?

2. Suppose that a deadly autosomal recessive disorder were to suddenly become dominant. What changes would you predict?

3. Draw a pedigree that demonstrates inheritance of an X-linked trait.

4. How common is polyploidy in humans?

5. Explain the connection between mitosis, nondisjunction, and cancer.

6. What is the difference between a translocation and a nondisjunction?

Multiple Choice

Circle the letter that best answers the question.

1. An individual with an autosomal recessive disorder generally has _____ copy/copies of the dominant allele.
   a. 0
   b. one
   c. two
   d. two, on the X chromosome
   e. two, on the Y chromosome

2. Which of the following is an autosomal disease?
   a. sickle-cell anemia
   b. red-green color blindness
   c. hemophilia
   d. Turner syndrome
   e. Klinefelter syndrome

3. X-linked disorders:
   a. are more common in females than males.
   b. never occur.
   c. are more common in males than females.

d. occur equally frequently in males and females.
   e. affect the Y chromosome.

4. An older mother may be at increased risk for having a child with which of the following disorders?
   a. Huntington disease
   b. Turner syndrome
   c. Down syndrome
   d. Klinefelter syndrome
   e. none of the above

5. Some flies have a diploid chromosome number of 8. Which of the following chromosome counts represents an aneuploid chromosome number?
   a. 0
   b. 4
   c. 8
   d. 9
   e. 16

6. Which of the following chromosome numbers would be found in a polyploid human embryo?
   a. 23
   b. 46
   c. 69
   d. 47
   e. none of the above

7. Which of the following is a syndrome that is more likely to be found in Reginald Pisum than in his wife, Pearl?
   a. Turner syndrome
   b. Klinefelter syndrome
   c. color blindness
   d. a and c
   e. b and c

8. Which of the following describes a situation in which the gene is not lost, but simply misplaced?
   a. translocation
   b. transcription
   c. transportation
   d. transcendental
   e. transformation

9. A new student joins your class. She is small of stature and does not seem to have developed sexually. You suspect the she might have:
   a. red-green color blindness.
   b. hemophilia.
   c. sickle-cell anemia.
   d. Turner syndrome.
   e. Huntington disease.

10. Which of the following is not a change in the structure of a chromosome?
    a. deletion
    b. translocation

c. inversion

d. nondisjunction

e. duplication

11. A mutation occurs in a gene that codes for black coat color in a rare species known as "crying" hyenas. Which of the following phenotypes is most likely to be the result of such a change?
    a. baldness
    b. four ears
    c. white coat
    d. no tail
    e. muteness (non-"crying")

12. Three rat embryos are generated during an in vitro fertilization experiment. Of the genotypes listed below, which would be male?
    a. XO
    b. XXY
    c. XXX
    d. a and b
    e. none of the above

13. Which of the following terms describes a failure of chromosomes to sort properly during meiosis?
    a. inversion
    b. translocation
    c. deletion
    d. duplication
    e. nondisjunction

14. Which of the following terms describes the exchange of material between non-homologous chromosomes?
    a. inversion
    b. translocation
    c. deletion
    d. duplication
    e. nondisjunction

Use the figure below to answer questions 15–16.

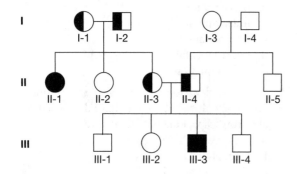

15. Based on the lineage depicted in the above figure, what can you conclude?
    a. the trait is dominant
    b. the trait is recessive

c. the trait is carried on the X chromosome

d. the trait is carried on the Y chromosome

e. there is insufficient evidence to conclude any of the above

16. Consider individual III-3. What can you predict about this individual?
    a. It is a male and has the trait of interest.
    b. It is a female and has the trait of interest.
    c. It is a male and is a carrier for the trait.
    d. It is a female and is a carrier for the trait.
    e. It does not carry the allele for the trait.

17. What term would you use to describe a portion of DNA that has been completely lost?
    a. inversion
    b. translocation
    c. deletion
    d. duplication
    e. nondisjunction

18. Your friend Becky tells you of a strange skin condition found in her family. Men on her father's side share a strange, camouflage-like melanin pattern on their backs. The condition is never seen in females of this lineage. What can you surmise about this condition?
    a. nothing
    b. It is unlikely to be genetic, because it appears only in some family members.
    c. It is likely to be genetic and autosomal.
    d. It is likely to be genetic and sex-linked.
    e. It is likely due to environmental factors.

19. If you were a carrier for sickle-cell anemia, which of the following mates would produce more malaria-safe children?
    a. an individual who is homozygous dominant for sickle-cell trait
    b. an individual who is homozygous recessive for sickle-cell trait
    c. an individual who is heterozygous for sickle-cell trait
    d. an individual who is either homozygous recessive or heterozygous
    e. an individual who is either homozygous dominant or heterozygous

20. Assuming that you chose the correct answer for question 19, what is the predicted highest proportion of malaria-safe children you could produce with this marriage?
    a. 0%
    b. 25%
    c. 50%
    d. 75%
    e. 100%

21. You discover a new organism that you name *madisonia*. You decide to do a genetic analysis on *madisonia* and find that you can line up matching (homologous) chromosomes as sets of three instead of two. What term describes this configuration?
    a. ploidy
    b. haploidy
    c. diploidy
    d. triploidy
    e. tetraploidy

22. Assuming that your specimen of *madisonia* is a fully functioning adult of the species, what can you conclude, based on this chromosome configuration?
    a. *Madisonia* could be a plant.
    b. *Madisonia* could be a mammal.
    c. *Madisonia* could be a bacterium.
    d. *Madisonia* could be a fish.
    e. This type of configuration is rather common, so no conclusion can be drawn about the organism.

23. Marfan syndrome is an autosomal dominant disorder. If a child has Marfan syndrome, what can you conclude about his parents?
    a. Both parents must be homozygous dominant.
    b. Both parents must be homozygous recessive.
    c. Both parents must be heterozygous.
    d. At least one parent must be homozygous recessive.
    e. At least one parent must also have Marfan syndrome.

24. Which of the following chromosome numbers represents a human with aneuploidy?
    a. 45
    b. 46
    c. 47
    d. a and b
    e. a and c

25. What condition is caused by being otherwise normal but having no copies of the Y chromosome?
    a. Down syndrome
    b. color blindness
    c. hemophilia
    d. femaleness
    e. maleness

WHAT'S IT ALL ABOUT?

Here's a question to help you pull together what you've learned so far using this text.

Question: Identical twins result when a single, fertilized egg splits into two cells during the earliest stages of development and each individual cell develops into an embryo. Despite a common set of genes, these twins still do not mature into indistinguishable people—although physically, mentally, and emotionally very similar, they are clearly different people. If the genotype drives the phenotype, why aren't identical twins identical people?

1. **What type of question is this?**
   Compare and contrast? Defend a position? Describe the effect of "A" on "B"?

2. **What kind of evidence do you need?**
   What kind of information do you need to provide to explain how genetically identical people can be different?

3. **Pull it all together.**
   Go for it!

CHAPTER 13 PASSING ON LIFE'S INFORMATION: DNA STRUCTURE AND REPLICATION

Basic Chapter Concepts

* Just as atoms are the basic units of matter, genes are the basic units of heredity. Genes carry information for the synthesis of proteins.
* The molecular structure of DNA, or the three-dimensional order of the atoms in space, was proposed in 1953 by James Watson and Francis Crick, based on the experimental data of Rosalind Franklin and Maurice Wilkins. Watson and Crick's model revolutionized the study of genetics and ushered in the field of molecular biology.
* The double helix of DNA consists of two repeated polynucleotide chains associated through hydrogen bonds between the nitrogen bases. The complementarity of the bases means that a DNA molecule can be a template for its own replication. The order of the bases along the chain constitutes a code for specifying the order of amino acids within a protein.
* Changes in genetic information, called mutations, occur when the sequence of nucleotide bases in the DNA changes.

CHAPTER SUMMARY

13.1 What Do Genes Do, and What Are They Made Of?
* The work of scientists during the 1930s and 1940s showed conclusively that genes are composed of DNA.
* The study of DNA initiated the science of molecular biology, the investigation of life at the level of individual molecules.

13.2 Watson and Crick: The Double Helix
* Determining how DNA could store and transmit genetic information required understanding the physical structure of the molecule.
* Watson and Crick's seminal double-helix model was inferred using experimental data from several research teams, particularly data from the X-ray crystallography experiments of Rosalind Franklin and Maurice Wilkins.

13.3 The Components of DNA and Their Arrangement
* DNA is composed of sugar molecules linked together by phosphate bonds. Each sugar molecule also has one of four nitrogen-containing molecules bound to it—adenine, thymine, cytosine, or guanine. These nitrogen bases can form hydrogen bonds with each other—adenine with thymine, and cytosine with guanine. This hydrogen bond can hold two DNA chains together such that the hydrogen-bonded nitrogen bases are on the inside of the helix and the sugar-phosphate bonds form the outside edges.
* The specificity of the hydrogen bonding between the bases explains the complementarity of adenine for thymine and cytosine for guanine, and it indicates how each DNA strand can serve as a template for the replication of the helix.
* An enzyme critical to DNA replication is DNA polymerase, which catalyzes the addition of new nucleotides to each of the DNA strands as they are unwound during replication.
* The order of the bases along the DNA strand serves as a code that specifies how the protein chain is to be assembled.

13.4   Mutations

- Mutation, a permanent change in the sequence of bases in a DNA molecule, may occur when bases are mispaired during replication. Subsequent rounds of mitosis preserve the change and pass it along to daughter cells.
- A mutation in the DNA of the germ-line cells, the gametes, allows the changed DNA to appear in every cell in the organism and passes the change from one generation to the next.
- Mutations are usually harmful, compromising the function of the protein specified by the DNA. On rare occasion, mutations may add new information to the genome and thus provide the raw material for evolutionary adaptations.

WORD ROOTS

nucleo- = nucleus (e.g., a *nucleo*tide is a single unit of nucleic acid)

mutat- = to change (e.g., a *mutat*ion is a permanent change in a cell's DNA)

soma- = body (e.g., a *soma*tic cell is any body cell that is not a germ cell)

KEY TERMS

DNA polymerase _____

germ-line cell _____

molecular biology _____

mutation _____

nucleotide _____

point mutation _____

somatic cell _____

FLASH CARDS

To use the flash cards, tear the page from the book and cut along the dotted lines. The key term appears on one side of the flash card, and its definition appears on the opposite side.

DNA polymerase	nucleotide
germ-line cell	point mutation
molecular biology	somatic cell
mutation	

the building block of nucleic acids, including DNA and RNA, consisting of a phosphate group, a sugar, and a nitrogen-containing base

an enzyme that is active in DNA replication, separating strands of DNA, bringing bases to the parental strands, and correcting errors by removing and replacing incorrect base pairs

a mutation of a single base pair in a genome

the succession of parent and daughter cells that ultimately produces either eggs or sperm

any cell that is not and will not become an egg or sperm cell

the investigation of life at the level of its individual molecules

a permanent alteration of a DNA base sequence

SELF TEST

Once you have finished studying this chapter, close your books, grab a pencil, and spend the next 15 to 20 minutes completing this practice test.

Compare and Contrast

For each of the following paired terms, write a sentence of comparison ("Both") and a sentence of contrast ("However,").

germ-line cell/somatic cell
mutation/point mutation
purine/pyrimidine
nucleotide/DNA
DNA polymerase/DNA replication

Short Answer

1. Prior to the 1930s, the concept of what a gene is and does was rather foggy. What happened in later years to clarify what a gene is and does?

2. Who did Watson and Crick collaborate with to elucidate the structure of DNA?

3. Is DNA replication completely error free? What measures does a cell have to deal with errors?

4. Will a child of a person with a mutation in a skin cell inherit that mutation? Explain.

Multiple Choice

Circle the letter that best answers the question.

1. How may mutations occur in a genome?
   a. through proofreading
   b. through uncorrected errors during DNA replication
   c. from temporary insertion of the wrong nucleotide in a new DNA strand
   d. none of the above
   e. all of the above

2. What is/are possible consequences of mutations?
   a. no DNA replication
   b. cancer
   c. evolution
   d. slower DNA replication
   e. the enzymes for DNA replication may not function

3. A mutation in a liver cell:
   a. will be inherited by the daughter cells of that liver cell.
   b. will cause immediate liver disease.
   c. will cause immediate death of the organism!
   d. will be inherited by the offspring of the organism with the mutation.
   e. will be passed on through meiosis.

4. The study of genetics at the level of DNA is known as:
   a. quantum mechanics.
   b. ethnobotany.
   c. computational biology.
   d. molecular biology.
   e. sociobiology.

5. Which of the following contains the actual code for building proteins?
   a. phosphate alignment
   b. deoxyribose bonds
   c. base sequences
   d. order of the chromosomes
   e. order of the deoxyribose molecules

6. Which of the following individuals is not credited as participating in the discovery of the double helix?
   a. Watson
   b. Mendel
   c. Crick
   d. Franklin
   e. Wilkins

7. Reginald Pisum was exposed to high doses of mutagenic compounds because of a bizarre industrial accident. Should he be concerned about having mutant children?
   a. No, mutations affect only adults.
   b. No, mutations have never been expressed in humans.
   c. Yes, if the mutation occurs in germ-line cells.
   d. Yes, if the mutation affects the autosomal cells.
   e. Yes, if the mutation affects his liver cells, the kids will have mutant livers, too.

8. An accident has occurred in the lab. The computer that you have been using to sequence the DNA of yellow-bellied sapsuckers has malfunctioned, and you fear that many months of data have been lost. However, you manage to recover part of a sequence of one strand of DNA. Is the gene for this sequence gone for good?
   a. Yes, you need to have both strands in order to know the gene.
   b. No, the gene may be on the "surviving" strand.
   c. Yes, you must have totally intact strands.
   d. No, you can recover any gene if you have any part of a DNA molecule.
   e. Yes, you should go back to the original organism and sequence it again.

Match the following terms with their description.
Each choice may be used once, more than once, or
not at all.

    a. deoxyribose
    b. adenine
    c. replication
    d. mutation
    e. polymerase

9. _____ A change in the gene sequence

10. _____ Process of creating "new" chromosomes

11. _____ One of the nitrogen-containing bases

12. _____ The sugar portion of a nucleotide

13. _____ The assembly and error-checking molecule

14. A gene is:
    a. the same thing as a chromosome.
    b. the information for making a specific protein.
    c. made of carbon molecules.
    d. made by a ribosome.
    e. made of protein.

15. A DNA molecule is a polymer made of:
    a. bases.
    b. amino acids.
    c. nucleotides.
    d. nucleic acids.
    e. sugar and phosphate.

16. Scientists studying melanoma cells found that the cells:
    a. all had a mutation in the cells called BRAF.
    b. all had a mutation in their DNA.
    c. may have a mutation in a gene called BRAF.
    d. grew abnormally large.
    e. produced a protein that caused them to grow abnormally slowly.

17. Huntington disease is a terrible genetic disease that causes progressive dementia and death to its victims. The basic defect in this disease is:
    a. production of abnormal amounts of the protein huntingtin.
    b. production of abnormal amounts of a faulty version of the protein huntingtin.
    c. due to an abnormally shaped chromosome 15.
    d. due to multiple copies of chromsome 15.
    e. due to multiple repeats of the sequence GTC.

18. In DNA, base pairing occurs between:
    a. cytosine and uracil.
    b. adenine and guanine.
    c. adenine and thymine.
    d. deoxyribose and phosphate.
    e. adenine and uracil.

19. Rosalind Franklin used which technique to determine some of the physical characteristics of DNA?
    a. light microscopy
    b. centrifugation
    c. X-ray diffraction
    d. spectroscopy
    e. electron microscopy

20. The DNA molecule could be compared to a:
    a. road with two lanes.
    b. ladder.
    c. lock and key.
    d. jigsaw puzzle.
    e. globular mass.

21. If a cell had an abnormal form of DNA polymerase in its nucleus, what would be the consequences?
    a. The cell would not be able to carry out accurate DNA replication.
    b. The daughter cells would have no DNA.
    c. The daughter cells would have no additional DNA.
    d. The cell would immediately die.
    e. The cell would mutate to a completely different kind of cell.

22. Which of the answers listed below are people who did *not* work in fields related to genetics?
    a. Mendel and Morgan
    b. Beadle and Tatum
    c. Wallace and Darwin
    d. Watson and Crick
    e. Franklin and Wilkins

23. Beadle and Tatum used *Neurospora* to study:
    a. mutations in fruit flies.
    b. the relationship between mutations and genes.
    c. the relationship between mutations and chromosomes.
    d. what *Neurospora* needs to grow the largest mold.
    e. what X rays do to the color of the mold.

24. The most common cause of mutations is:
    a. random errors in DNA replication.
    b. exposure to X rays.
    c. exposure to secondhand cigarette smoke.
    d. exposure to UV radiation.
    e. exposure to car exhaust.

25. Which of the answers listed below are *not* correctly paired?
    a. A-C
    b. C-G
    c. A-T
    d. T-A
    e. G-C

WHAT'S IT ALL ABOUT?

Here's a question to help you pull together what you've learned so far using this text.

Question: In this chapter, we have seen how a single nucleotide change in the DNA, a point mutation, can be propagated during cell division thanks to the efficiency of DNA replication. Because mutations occur spontaneously during DNA replication, mutations can occur in gamete-producing cells as well as the cells of other tissues. Why, then, do heritable mutations that produce a unique phenotype appear only rarely, but mutations that cause cancer, for example, seem more common?

1. **What type of question is this?**
   Compare and contrast? Defend a position? Describe the effect of "A" on "B"?

2. **What kind of evidence do you need?**
   What do you know about DNA replication, mitosis, and gametogenesis? All of these processes factor into this question. What do you know about mutations, genotypes, and effect on phenotype? How can mutations in the genotype produce a phenotype (cancer)?

3. **Pull it all together.**
   Go ahead, exercise those brain cells!

CHAPTER 14 HOW PROTEINS ARE MADE: GENETIC TRANSCRIPTION, TRANSLATION, AND REGULATION

Basic Chapter Concepts

- Proteins are essential to life. They provide structure and catalyze metabolic reactions.
- The first step in the process of converting genetic information into protein requires DNA to be transcribed into a form that can carry information out to the protein-synthesis machinery in the cytoplasm. This "messenger" is a molecule of RNA, aptly named messenger RNA (mRNA).
- The second step in the process translates the code held in the mRNA into a sequence of amino acids. The code uses three nucleotide bases to specify a single amino acid, hence the name "triplet code." The code is also redundant; an amino acid may be specified by more than one triplet of nucleotide bases.
- Cells use various signals to regulate protein synthesis; specific sequences within the DNA recognize the signal molecules and determine the rate of protein synthesis.

CHAPTER SUMMARY

14.1 The Structure of Proteins
- Proteins are composed of a series of amino acid subunits. There are 20 types of amino acids, so proteins can have almost infinite variety of structure.

14.2 Protein Synthesis in Overview
- The process of converting a message encoded in DNA into protein requires two stages: transcription and translation.

14.3 A Closer Look at Transcription
- RNA, ribonucleic acid, is also a polymer composed of ribose sugars holding nitrogen bases and held together by phosphate bonds. It is not a double helix; RNA functions as a single-stranded molecule, and it does not contain thymine, replacing it with the base uracil.
- Complementary base pairing allows an RNA molecule to be synthesized using DNA as a template, allowing the transfer of information from DNA into this mRNA transcript.

14.4 A Closer Look at Translation
- The mRNA leaves the nucleus and binds to the ribosome, a cytoplasmic organelle where all the molecules needed to make proteins are assembled.
- The mRNA sequence is translated into protein through the action of transfer RNA (tRNA), a small RNA molecule with two functional sites. One site can hold a specific amino acid while the second site "reads" the triplet code of nitrogen bases within the RNA.

14.5 Genetic Regulation
- DNA contains information, but only when proteins act on that information can DNA function in the cell. However, controlling whether the DNA may be acted on is an effective way to control protein translation.

14.6 Genetics and Life
- There is no life without genetic material.

WORD ROOTS

poly- = many (e.g., a *poly*merase is an enzyme that as-
sembles many subunits)

trans- = across (e.g., *trans*lation converts the molecular message from one language [nucleic acid] to
another [protein])

KEY TERMS

alternative splicing _____

_____

anticodon _____

_____

codon _____

_____

genetic code _____

_____

messenger RNA (mRNA) _____

_____

polypeptide _____

_____

promoter sequence _____

_____

ribosomal RNA (rRNA) _____

_____

ribosome _____

_____

RNA polymerase _____

_____

transcription _____

_____

transfer RNA (tRNA) _____

_____

translation _____

_____

FLASH CARDS

To use the flash cards, tear the page from the book and cut along the dashed lines. The key term appears on one side of the flash card, and its definition appears on the opposite side.

alternative splicing	ribosomal RNA (rRNA)
anticodon	ribosome
codon	RNA polymerase
genetic code	transcription
messenger RNA (mRNA)	transfer RNA (tRNA)
polypeptide	translation
promoter sequence	

a type of RNA that, along with proteins, forms ribosomes

a process in genetics in which a single primary transcript can be edited in different ways to yield multiple messenger RNAs, which in turn yield multiple proteins

an organelle, located in the cell's cytoplasm, that is the site of protein synthesis; the translation phase of protein synthesis takes place within ribosomes

the end of the transfer RNA molecule that can bind with a particular codon on the mRNA transcript

in the transcription phase of protein synthesis, the enzyme that unwinds the DNA double helix and puts together a chain of RNA nucleotides complementary to the exposed DNA nucleotides

an mRNA triplet that codes for a single amino acid or a start or a stop command in the translation stage of protein synthesis

in protein synthesis, the process in which DNA's information is copied onto messenger RNA (mRNA)

the inventory of linkages between nucleotide triplets and the amino acids they code for; with few exceptions, the genetic code is universal in living things

in protein synthesis, a form of RNA that bonds with amino acids, transfers them to ribosomes, and then bonds with a messenger RNA sequence

a type of RNA that encodes, and carries to ribosomes, information for the synthesis of proteins

the process in which a polypeptide chain is produced within a ribosome based on the information encoded in messenger RNA; this process, the second major stage in protein synthesis (after transcription), occurs in the cell's cytoplasm

a series of amino acids linked in linear fashion; polypeptide chains fold up to become proteins

the site on a segment of DNA to which RNA polymerase attaches prior to beginning transcription

SELF TEST

After you have finished studying this chapter, close your books, grab a pencil, and spend the next 15 to 20 minutes working on this practice test.

Compare and Contrast

For each of the following paired terms, write a sentence of comparison ("Both") and a sentence of contrast ("However,").

transcription/translation
tRNA/mRNA
operator/repressor
intron/exon
codon/genetic code

Short Answer

1. Explain the significance of the Jacob-Monod model of the lac operon.

2. How is it that so many different types of proteins can be produced from only 20 amino acids?

3. Explain the impact of a mutation that affects the anticodon.

4. List the structural and functional differences between DNA and RNA.

5. Explain the impact of the discovery of micro-RNA molecules on our understanding of molecular biology.

6. "Life is made possible by the fantastic ability of genetic systems to store, use, and pass on information" (p. 265 in the text). Explain this statement in your own words.

Multiple Choice

Circle the letter that best answers the question.

1. How many mRNA bases make up a codon?
   a. 1
   b. 3
   c. 6
   d. 20
   e. varies, depending on the species

2. What portion of mRNA is removed during processing?
   a. extran
   b. exon
   c. intron
   d. promoter
   e. codons

3. How do tRNAs bring the correct amino acid to the growing polypeptide?
   a. by base pairing with the rRNA
   b. by complementary base pairing with the gene in the nucleus

   c. by covalently binding to the ribosome
   d. by complementary base pairing between the tRNA anticodon and the mRNA codon
   e. none of the above

4. Ribosomes are made up of:
   a. protein.
   b. RNA.
   c. lipids.
   d. a, b, and c
   e. a and b

5. Translation of a specific mRNA molecule usually occurs:
   a. with multiple ribosomes simultaneously translating.
   b. with many exons loading tRNAs on the codon.
   c. with only one ribosome at a time translating.
   d. with multiple RNA polymerases reading the same gene simultaneously.
   e. with both RNA and DNA nucleotides.

6. What is the correct flow of information from gene to protein?
   a. mRNA-gene-protein
   b. protein-gene-mRNA
   c. gene-protein-tRNA
   d. gene-rRNA-protein
   e. gene-mRNA-protein

7. The end of translation occurs when:
   a. an enzyme is sent from the DNA in the nucleus.
   b. endoplasmic substances attach to the ribosome.
   c. a termination codon in the mRNA is reached.
   d. a telo-tRNA anticodon attaches to the ribosome.
   e. the translator is disassembled.

8. Transcription occurs in the _____ while translation occurs in the _____.
   a. nucleus; nucleus
   b. endoplasmic reticulum; nucleus
   c. nucleus; cell membrane
   d. nucleus; cytoplasm
   e. cytoplasm; cytoplasm

9. An inducible gene is:
   a. turned on by DNA nucleotides.
   b. turned on by the presence of its substrate.
   c. turned on at random.
   d. found only in plants.
   e. not found in nature; it is an artificial construct.

10. RNA polymerase:
    a. builds transcripts.
    b. excises introns.
    c. brings amino acids to the mRNA.
    d. forms part of the ribosome.
    e. unwinds the double helix.

11. Reginald Pisum is feeling ill, with all the classic symptoms of a head cold. He may even have to stay home instead of attending the Truck & Tractor Pull Semifinals on Saturday night. How is it that one little virus can cause so much misery?
   a. Viruses are cells, and they crowd out the other cells in the body, causing illness.
   b. Viruses convert the protein-synthesis machinery within cells to make many copies of themselves, thereby spreading the infection.
   c. Viruses destroy cellular DNA.
   d. Viruses use the lac operon to control Reginald's body cells.
   e. Viruses don't contain thymine, so they interfere with protein synthesis.

12. You go home for Thanksgiving and mention that you have been learning about genetics in biology class. Great Aunt Michele asks you, "What *is* a gene anyway?" You respond:
   a. "A gene is a molecule that oversees the assembly of amino acids into a polypeptide."
   b. "A gene is a piece of RNA that makes DNA."
   c. "A gene is a segment of DNA that causes the transcription of a segment of RNA."
   d. "A gene is half of a pair of homologous chromosomes."
   e. "A gene is an enzyme that causes cell division."

13. Which of the following is a substance that induces the transcription of a gene?
   a. rRNA
   b. promoter
   c. anticodon
   d. transcript
   e. intron

14. Which of the following is a noncoding segment of DNA?
   a. rRNA
   b. protein
   c. anticodon
   d. promoter
   e. intron

15. _____ forms part of the ribosome.
   a. rRNA
   b. Lactose
   c. Anticodon
   d. Transcript
   e. Intron

16. Which of the following terms describes mRNA as it leaves the DNA?
   a. rRNA
   b. translater
   c. anticodon
   d. transcript
   e. intron

17. Which of the following is part of tRNA?
   a. rRNA
   b. lactose
   c. anticodon
   d. transcript
   e. intron

18. According to complemetary base-pairing rules, for every G there must be:
   a. T.
   b. A.
   c. C.
   d. G.
   e. either C or G.

19. Using the DNA sequence TACGGTACCATTGCGCAA, determine which of the following choices is the matching RNA molecule.
   a. AUGCCAUGGGAACGCGUU
   b. AUGCCUAGGUAACGCGUU
   c. AUGCCAUGGUAACGCCUU
   d. AUGGGAUGGUAACGCGUU
   e. AUGCCAUGGUAACGCGUU

20. Divide the correct mRNA sequence from Question 19 into codons. How many do you have?
   a. 2
   b. 4
   c. 6
   d. 12
   e. 18

21. Approximately how many amino acids does the sequence from Question 19 code for?
   a. two
   b. four
   c. six
   d. eight
   e. nine

22. Humans probably have no more than _____ genes.
   a. 10,000
   b. 15,000
   c. 20,000
   d. 25,000
   e. 30,000

23. What enzyme is responsible for transcription?
   a. DNA polymerase
   b. RNA poylmerase
   c. ligase
   d. transcriptase
   e. cyclase

24. One gene can produce several structurally different proteins through a process known as:
   a. alternative splicing.
   b. transcriptional editing.
   c. semiconservative replication.
   d. promotion.
   e. translocation.

25. A ribosome is:
   a. an organelle.
   b. a structural protein.
   c. an enzyme.
   d. RNA.
   e. DNA.

WHAT'S IT ALL ABOUT?

Here's a question to help you pull together what you've learned so far using this text.

Question: Every living thing (and nonliving virus) uses the same triplet code to translate mRNA into proteins. Interestingly, organisms show bias for certain codons; bacteria might prefer to use GCU to code for alanine, whereas plants might use GCA. What might account for this codon bias?

What do I do now?

Remember the drill—decide what the question is asking you to do, collect your evidence from this chapter (and the others you've studied) and write!

CHAPTER 15   THE FUTURE ISN'T WHAT IT USED TO BE: BIOTECHNOLOGY

Basic Chapter Concepts

- The tools of biotechnology allow us to produce large amounts of DNA from a small starting sample and to move DNA between organisms.
- Recombinant DNA cloning makes it possible to produce large quantities of pharmaceutically important human biomolecules, such as insulin or growth hormone. The availability of some hormones could lead to abuse.
- The potential use of biotechnology to modify our own species, either through reproductive cloning or recombinant DNA cloning, is an unresolved issue. Just because we can modify organisms, does that mean we should?
- Biotechnology raises many hotly debated ethical issues.

CHAPTER SUMMARY

15.1   What Is Biotechnology?
- Biotechnology is the use of technology to manipulate biological processes to meet societal needs.

15.2   Transgenic Biotechnology
- Restriction enzymes cut DNA at specific, usually asymmetrical, sequences. Pieces of DNA cut by the same enzyme can be annealed because they share complementary ends.
- Plasmids and bacteriophages are used as carriers of DNA sequences. DNA from any organism can be cloned into these plasmid or bacteriophage vectors and used to transform a population of bacteria to produce the protein product of the cloned DNA.
- Cloning facilitates the production of transgenic plants, crops altered to increase disease and pest resistance or to yield greater nutritive value. The effect of these plants on wild versions or beneficial insect species is uncertain.

15.3   Reproductive Cloning
- Entire organisms can be cloned through the process of reproductive cloning, in which the nucleus from any cell of one organism, such as a sheep, can be introduced into an enucleated egg cell of the same species using a small electrical current to simulate fertilization. The resulting embryo will be a genetic copy of the organism that donated the DNA (nucleus).

15.4   Forensic Biotechnology
- Copies of DNA sequences can be made using a small amount of DNA to serve as the template in the polymerase chain reaction (PCR). PCR increases the amount of DNA in a sample so that other analyses can be performed, and it has proved to be an invaluable technique for criminal investigations, disease diagnosis, and evolutionary studies.
- DNA profiles are based on short tandem repeats (STR), which are multiple repeats of specific DNA sequences that occur randomly in human genomes; by comparing the positions of the STR between victim(s) and suspect(s), forensic scientists can accurately show whether a certain individual was present at a crime.

15.5   Stem Cells
- Stem cells are unique because they are not yet programmed to become a specific cell type; they are adaptable and can give rise to a wide variety of cells.

- The most common source of stem cells is embryonic cells; these cells are usually harvested from the blastocyst, which is an early stage of embryonic development. However, it has been shown that adults also contain stem cells, in smaller numbers, in many tissues of the body.
- In the United States, federal funding for stem cell research has been sharply curtailed by order of President Bush. Therefore, some states and private industries are funding their own stem cell research.
- Cells derived from embryonic stem cells have the potential to treat disease and injury and have been shown to be particularly important in treating diseases of the nervous system and traumatic brain injuries.

15.6 Biotechnology in the Real World

- In the twenty-first century, biotechnology dominates the headlines in terms of animal cloning, genetically modified organisms, and gene therapy. However, progress in biotechnology is both costly and slow. Although genetically modified plants are routinely used in farming in the United States, Canada, and Mexico, there is resistance to genetically modified organisms among some countries of the European Union. Many countries, including the United States have established government commissions to study and educate the public about the possible benefits and misuse of biotechnology.

WORD ROOTS

bio- = life, living organisms or tissue (e.g., *bio*technology is the use of technology to control biological processes)

-cyst = sac (e.g., a blasto*cyst* is a hollow, fluid-filled ball of cells that is formed in the early stages of the embryonic development of humans and other mammals)

embryo = (e.g., *embryo*nic stem cells are cells from the blastocyst stage of a human embryo that is capable of giving rise to all the types of cells in the adult body)

-genic = producing, forming (e.g., a trans*genic* organism is an organism whose genome has stably incorporated one or more genes from another species)

KEY TERMS

biotechnology _____

_____

blastocyst _____

clone _____

cloning vector _____

_____

embryonic stem cells _____

_____

plasmid _____

_____

polymerase chain reaction (PCR) _____

_____

recombinant DNA _____

reproductive cloning _____

restriction enzyme _____

_____

somatic cell nuclear transfer (SCNT) _____

_____

therapeutic cloning _____

_____

transformation _____

_____

transgenic organism _____

_____

FLASH CARDS

To use the flash cards, tear the page from the book and cut along the dashed lines. The key term appears on one side of the flash card, and its definition appears on the opposite side.

biotechnology	recombinant DNA
blastocyst	reproductive cloning
clone	restriction enzyme
cloning vector	somatic cell nuclear transfer (SCNT)
embryonic stem cells	therapeutic cloning
plasmid	transformation
polymerase chain reaction (PCR)	transgenic organism

two or more segments of DNA that have been combined by humans into a sequence that does not exist in nature

the use of technology to control biological processes as a means of meeting societal needs

cloning intended to produce adult mammals of a defined genotype

hollow, fluid-filled ball of cells that is formed in the early stages of the embryonic development of humans and other mammals; in non-mammalian animals, the blastocyst is known as the blastula

a type of enzyme, occurring naturally in bacteria, that recognizes a specific sequence of DNA bases and cuts DNA strands at a specific location within the sequence; restriction enzymes are used in biotechnology to cut DNA in specific places

an exact genetic copy; also—used as a verb—to make one of these copies; a single gene or a whole, complex organism can be cloned

a means of cloning animals through fusion of one somatic (non-sex) cell with an egg cell whose nucleus has been removed (an "enucleated" cell)

a self-replicating agent that, in the cloning process, serves to transfer genetic material; examples include bacterial plasmids and the viruses known as bacteriophages

the use of cloning to produce human embryonic stem cells that can be used to treat disease

cells from the blastocyst stage of a human embryo that is capable of giving rise to all the types of cells in the adult body

a cell's incorporation of genetic material from outside its boundary; some bacteria readily undergo this process, and others can be induced to for uses in biotechnology

a ring of DNA that lies outside the chromosome in bacteria; plasmids can move into bacterial cells in the process called transformation, thus making them a valuable tool in biotechnology

an organism whose genome has stably incorporated one or more genes from another species

a technique for generating many copies of a DNA sequence from a small starting sample

SELF TEST

Once you have finished studying this chapter, close your books, grab a pencil, and spend the next 15 to 20 minutes completing this practice test.

Compare and Contrast

For each of the following paired terms, write a sentence of comparison ("Both") and a sentence of contrast ("However,").

reproductive cloning/therapeutic cloning
cloning vector/plasmid
blastocyst/embryonic stem cells
recombinant DNA/transformation
DNA profiling/short tandem repeats (STR)

Short Answer

1. How can a human protein be made by a bacterium?

2. What do restriction enzymes do?

3. What is cloning? How would you clone your best friend or dog?

4. What do you think the impact of PCR has been on the field of forensics?

5. What are the pros and cons of using embryonic stem cells versus adult stem cells?

6. Why do you think people distrust genetically modified organisms?

Multiple Choice

Circle the letter that best answers the question.

1. Which molecules are analogous to DNA scissors?
   a. plasmids
   b. vectors
   c. restriction enzymes
   d. human growth hormone
   e. none of the above

2. Dolly the sheep is:
   a. an exact copy of her father.
   b. an animal with DNA from two different cells in each cell of her body.
   c. a reproductive clone.
   d. haploid.
   e. c and d

3. You have three copies of a particular DNA molecule. What technique would you use to make more copies of the molecule?
   a. gel electrophoresis
   b. sequencing
   c. PCR

d. restriction fragment analysis
   e. none of the above

4. Recombinant DNA:
   a. requires the use of PCR to make it.
   b. is harvested from viruses.
   c. contains no thymine bases.
   d. demonstrates the universality of the genetic code.
   e. naturally occurs in most plants.

5. Biotechnology:
   a. is an ancient science.
   b. has been producing mammal clones for about 50 years.
   c. can cure every illness.
   d. produces only drugs.
   e. is so new that regulation has not caught up with it yet.

6. Bacterial plasmids are of particular interest to biotechnologists because:
   a. plasmids cause random mutations within human and other eukaryotic cells.
   b. plasmids glow spontaneously when used in goat cells.
   c. plasmids make the DNA of their cells rigid like plastic and easy to identify.
   d. plasmids can deliver DNA to other cells by the process of transformation.
   e. plasmids can deliver RNA to other cells by the process of transformation.

7. Restriction enzymes are:
   a. bacterial enzymes that can destroy viruses.
   b. bacterial enzymes that destroy other bacteria.
   c. enzymes that are capable of self-destruction.
   d. used only in forensic studies.
   e. viral enzymes that can destroy bacteria.

8. What is a plasmid?
   a. an organelle found in bacteria
   b. an organelle found in plants
   c. a circular piece of bacterial DNA that copies independent of the main bacterial chromosome
   d. a type of virus that has a nucleus
   e. a type of bacterium that has a nucleus

9. If an organism is transgenic, then:
   a. it is no longer safe to consume.
   b. it has one or more genes from another organism.
   c. it is no longer the same species.
   d. it has 50 percent of its own genes and 50 percent from another species.
   e. it no longer has the same number of chromosomes.

10. During the process of _____, _____ is incorporated from one cell to another.
    a. replication; DNA
    b. transformation; a virus
    c. mitosis; DNA
    d. meiosis; DNA
    e. transformation; DNA

11. Recombinant DNA technology can be used to:
    a. show that all organisms have the same DNA.
    b. insert genes into bacteria to produce specific proteins.
    c. insert genes into bacteria to produce more DNA.
    d. insert new nuclei into bacteria.
    e. combine multiple genomes to produce new organisms.

12. Which of the organisms listed below are *not* commonly used in transgenic biotechnology?
    a. bacteria
    b. yeast
    c. mice
    d. goats
    e. earthworms

13. Food containing genetically modified organisms (GMO) could be on your table tonight. This means that the food is:
    a. enriched with extra vitamins.
    b. treated with pesticides.
    c. genetically altered.
    d. probably more expensive than non-GMO food.
    e. more likely to make you ill.

14. The polymerase chain reaction (PCR) is an important technique in biotechnology because it allows researchers to:
    a. insert bacterial genes into plasmids.
    b. make transgenic organisms.
    c. make many copies of a specific DNA sequence.
    d. insert viral genes into plasmids.
    e. bioengineer total organisms from tiny DNA fragments.

15. Although both plants and animals can be genetically modified, researchers have found that genetically engineered plants:
    a. are more difficult to grow.
    b. can be grown only in the United States and not in other countries.
    c. can be enriched with carotene, as in golden rice.
    d. do not need to be fertilized.
    e. can produce proteins that attract insects.

16. Every species has its own distinct genome. How is it possible to transfer genes from one species to another?
    a. All organisms have similar cell structure.
    b. Even if the genes differ, all organisms have the same cells.
    c. Genes produce specific proteins, no matter which organism they are found in.
    d. All organisms are subject to mutations.
    e. All organisms have DNA.

17. The difference between stem cells and other body cells is that:
    a. stem cells can be obtained only from embryos.
    b. stem cells are adaptable and can give rise to other cell types.
    c. stem cells can be used to create entire new species.
    d. body cells usually do not have nuclei.
    e. stem cells are difficult to culture in the lab, whereas body cells grow at a very fast rate.

Match the correct term with its definition.
    a. polymerase chain reaction
    b. primer
    c. plasmid
    d. somatic cell nuclear transfer
    e. restriction enzyme

18. _____ A molecule used to carry foreign genes into bacteria

19. _____ A rapid way to amplify DNA in the laboratory

20. _____ Molecular scissors

21. _____ Method for cloning mammals

22. _____ Short DNA segments that act as signals to DNA polymerase in the PCR reaction

23. Human genomes commonly have short DNA sequences that are repeated from 3 to 50 times. These are known as:
    a. codons.
    b. genetic stutters.
    c. short tandem repeats.
    d. multiple gene copies.
    e. mutations.

24. Until about 20 years ago, law enforcement agencies primarily used fingerprints and blood as forensic evidence. However, forensic DNA typing now allows law enforcement officers to:
    a. compare DNA from victims with possible suspects.
    b. identify victims from terrorist attacks where only body fragments are found.

c. compare short tandem repeat patterns among suspects and victims.

d. exonerate people who have been wrongly convicted of crimes.

e. all of the above

25. Researchers have now found that some adult cells can be used as stem cells. However, adult stem cells are not as successful as embryonic stem cells because:

a. not all adults have the same kinds of cells.

b. every type of adult cell differs in its genetic makeup.

c. adult stem cells are limited in their ability to differentiate into other cells.

d. adult stem cells are found only in muscles and blood.

e. adult stem cells have more mutations than do embryonic stem cells.

WHAT'S IT ALL ABOUT?

Here's a question to help you pull together what you've learned so far using this text.

Question: Imagine that it is September 2001 and that you are employed by the Coroner's Office in New York City as a forensic technician. You are surveying the site of one of the most devastating mass disasters our country has ever witnessed, and your job is to officially identify the victims so that insurance claims and other legal processes can be settled. What are you going to do to certify death certificates for thousands of people whose remains are fragmentary?

What do I do now?

Remember the drill—decide what the question is asking you to do, collect your evidence from this chapter (and the others you've studied), and write!

CHAPTER 16 AN INTRODUCTION TO EVOLUTION: CHARLES DARWIN, EVOLUTIONARY THOUGHT, AND THE EVIDENCE FOR EVOLUTION

Basic Chapter Concepts

- All living things on Earth have descended from a common ancestor; modifications of ancestral forms over extraordinarily long time periods have produced the great variety of living species now in existence.
- Living things, humans included, are neither static nor directed on a particular course of development.
- Natural selection drives evolution; that is, the interaction of the living organism with the nonliving environment drives change blindly.
- Six independent lines of evidence support Darwin's theory of descent with modification, the cornerstone of evolutionary theory.

CHAPTER SUMMARY

16.1 Evolution and Its Core Principles
- Living organisms possess a suite of traits and behaviors that determines their survival and reproductive success. Collections of traits that enhance reproductive success allow an organism to leave more surviving offspring, who pass those traits on to future generations.

16.2 Charles Darwin and the Theory of Evolution
- Although Charles Darwin was not the only person to describe the process of descent with modification, he provided observational evidence supporting the hypothesis of common descent with modification and first described the selection pressure exerted by natural forces.

16.3 Evolutionary Thinking before Darwin
- Darwin's thinking about the origin of species was influenced by Charles Lyell's geological observations suggesting that the Earth itself was not static and by Georges Cuvier's fossil evidence.

16.4 Darwin's Insights Following the *Beagle*'s Voyage
- After returning from the *Beagle* voyages, Darwin's ideas were refined by reading T. R. Malthus's writings about human populations.

16.5 Alfred Russel Wallace
- Another Englishman, Alfred Russel Wallace, developed a theory about the effect of natural selection on populations.

16.6 Descent with Modification Is Accepted
- In 1859, Darwin published his theories in *On the Origin of Species by Means of Natural Selection*.

16.7 Darwin Doubted: The Controversy over Natural Selection
- After the turn of the twentieth century, new data from other areas of science helped to demonstrate that natural selection is the most important process underlying evolution. Data from genetics, the fossil record, and biogeography all support the significance of natural selection in evolution.

16.8 Opposition to the Theory of Evolution

- Resistance to the principles of Darwin's work may stem from a misunderstanding of scientific nomenclature. To a scientist, a theory is not a guess but instead is supported by available evidence.

16.9 The Evidence for Evolution

- Radiometric dating, which uses the "clock" of radioactive decay to measure time, indicates that the Earth is about 4.6 billion years old—old enough to provide the time for evolution.
- Fossil evidence shows the change from "simple" organisms in old layers of rock to more complex organisms in more recent layers, as the theory of evolution would predict.
- Embryology (the study of animal development) and morphology show that different-looking structures in adult animals develop from common structures in the embryo, as we would expect if animals had developed from a common ancestor.
- All organisms share a common genetic code, which is strong evidence for a common ancestor.
- The predictions of evolutionary theory are borne out in laboratory experiments using populations of rapidly reproducing organisms, such as fish and bacteria.

WORD ROOTS

homo- = same (e.g., structures of similar origin are described as *homo*logous)

morph- = shape (e.g., *morph*ology is the study of animal structures)

KEY TERMS

biogeography _____

common descent with modification _____

embryology _____

homologous _____

modern synthesis _____

morphology _____

natural selection _____

radiometric dating _____

FLASH CARDS

To use the flash cards, tear the page from the book and cut along the dashed lines. The key term appears on one side of the flash card, and its definition appears on the opposite side.

biogeography	modern synthesis
common descent with modification	morphology
embryology	natural selection
homologous	radiometric dating

the unified evolutionary theory that resulted from the convergence of several lines of biological research between 1937 and 1950

the geographic distribution of living things

the physical form of an organism

the process by which species of living things undergo modification in successive generations, with such modification sometimes resulting in the formation of new, separate species

a process in which the differential adaptation of organisms to their environment selects those traits that will be passed on with greater frequency from one generation to the next

the study of how animals develop from fertilization to birth

a technique for determining the age of objects by measuring the decay of the radioactive elements within them; the age of fossils can be determined with this technique

in anatomy, having the same structure owing to inheritance from a common ancestor; forelimb structures in whales, bats, cats, and gorillas are homologous

SELF TEST

After you have finished studying this chapter, close your books, grab a pencil, and spend the next 15 to 20 minutes completing this practice test.

Compare/Contrast

For each of the following paired terms, write a sentence of comparison ("Both") and a sentence of contrast ("However,").

paleontology/taxonomy
natural selection/fitness
embryology/homologous structures
evolution/natural selection
Darwin/Wallace

Short Answer

1. Explain descent with modification.
2. Who was Malthus? How did he influence Darwin?
3. Who was Lamarck?
4. Where did Darwin gather the evidence that he used to define his theory of evolution?
5. Who was Alfred Russel Wallace?
6. What work by Darwin was published in 1859?
7. How did the modern synthesis help natural selection gain acceptance?
8. Explain what is meant by the term *scientific theory*.
9. What is meant by the term *descent with modification*?

Multiple Choice

Circle the letter that best answers the question.

1. We would predict that two very distantly related species, such as bluebirds and alligators, probably:
   a. share most of the same DNA sequences.
   b. share more DNA sequences than a bluebird and a duck.
   c. share relatively few DNA sequences.
   d. share all the same DNA sequences.
   e. share no DNA sequences.

2. Is the layering of fossils in the Earth used to support evolution?
   a. Yes, fossils are laid down in chronological order, and therefore we can observe changes in these species over time.
   b. No, the fossil evidence is too recent to provide support.
   c. No, because Darwin didn't use it, the modern synthesis also ignores it.
   d. Yes, the fossil evidence is the only support for Darwin's idea of acquired characteristics.
   e. Yes, the fossil record provides evidence for how present-day (extant) species have changed their embryonic developmental patterns in the past 100 years.

3. You take a job as a research assistant at one of the richest fossil fields ever to be discovered. On your first day, you discover a new species, *Neko borgus,* that bears similarities to the modern domesticated cat. The bed where this fossil was found also contained several fossils of early primates. From this you conclude that:
   a. *Neko borgus* must have been active millions of years before the primates appeared.
   b. *Neko borgus* must have been active millions of years after the primates appeared.
   c. *Neko borgus* must have been active at about the same time as the primates appeared.
   d. the mixing of fossils was likely due to geological events like those described by Charles Lyell.
   e. early primates kept domesticated cats.

4. What is the supportive evidence for evolution from comparative anatomy?
   a. All antelopes have four legs.
   b. Most animals have a head.
   c. Closely related species have similar structures.
   d. Similar gene sequences code for similar proteins.
   e. Diverse species have homologous structures.

5. What is the supportive evidence for evolution from comparative embryology?
   a. All plant seeds look alike.
   b. All embryos arise from the union of egg and sperm.
   c. Different species have different embryos.
   d. Different species develop along the pattern set by their common ancestor.
   e. All baby animals look alike.

6. Which of the following provide(s) evidence for the process of evolution?
   a. radiometric dating
   b. the fossil record
   c. molecular biology
   d. genetic studies
   e. all of the above

7. What is a consequence of natural selection?
   a. Species become extinct.
   b. Individuals that better fit the environment leave fewer offspring.

c. Certain species will be preserved in the fossil record.

d. All organisms share a common ancestor.

e. Stronger individuals are more likely to survive.

8. How many life forms must have existed when life originated on Earth?

a. as many as every living species today

b. as many as every species that is living and has ever lived on Earth

c. one

d. three—a bacterial ancestor, a plant ancestor, and an animal ancestor

e. approximately the population of the Earth today

9. Which of the following scientists is *not* correctly matched with his theory?

a. Cuvier—continuous fossil record with no apparent breaks

b. Darwin—natural selection as a mechanism for evolution

c. Wallace—natural selection as a mechanism for evolution

d. Lamarck—characters acquired during an organism's lifetime are passed to offspring

e. Lyell—dynamic geological events continue to shape the Earth

10. Which of the following are examples of evolution?

a. different finch species living on different Galapagos islands

b. the rise of antibiotic-resistant strains of bacteria

c. changes in guppy populations after the introduction of predators

d. all of the above

e. none of the above

11. How does the modern synthesis differ from Darwin's theory of evolution?

a. The modern synthesis includes the evidence from several lines of scientific research.

b. The modern synthesis was first proposed by Alfred Russel Wallace.

c. The modern synthesis was introduced by Watson and Crick after the discovery of DNA structure.

d. The modern synthesis was the term Darwin first used to describe descent with modification.

e. The modern synthesis incorporates both intelligent design and natural selection.

12. Which of these would *not* be homologous to the others?

a. arm bones of a human

b. front leg bones of a dog

c. front fins of a fish

d. bones of a dolphin flipper

e. turtle leg bones

13. Evolution is a:

a. concept.

b. hypothesis.

c. belief.

d. theory.

e. guess.

14. Darwin's work was based on his observations in:

a. the Galapagos Islands.

b. the Caribbean.

c. the Urals.

d. Indonesia.

e. the Sinai.

15. Pharyngeal slits become:

a. eustachian tubes in humans.

b. feathers in birds.

c. gills in fish.

d. mouth parts in insects.

e. a and c

16. The alternative model to Mendelian genetics that explains inheritance is:

a. meiotic division.

b. descent with modification.

c. mitotic inheritance.

d. holistic division.

e. blending inheritance.

17. Comparing structures from two different species is the science of:

a. taxonomy.

b. morphology.

c. embryology.

d. a and b

e. all of the above

18. _____ proposed that species evolved by the inheritance of characters acquired by their parents.

a. Darwin

b. Lamarck

c. Lyell

d. Malthus

e. Wallace

19. _____ proposed that geological processes are still ongoing.

a. Darwin

b. Lamarck

c. Lyell

d. Malthus

e. Wallace

20. _____ provided the "icing on the cake" for Darwin's theory.

a. Darwin

b. Lamarck

c. Lyell

d. Malthus

e. Wallace

21. _____ wrote a letter proposing natural selection and common descent with modification as the mechanism for evolution.
 a. Darwin
 b. Lamarck
 c. Lyell
 d. Malthus
 e. Wallace

22. _____ sailed on the HMS *Beagle.*
 a. Darwin
 b. Lamarck
 c. Lyell
 d. Malthus
 e. Wallace

23. Given the following timeline, where would you place these events?
 i. Darwin attends Cambridge
 ii. Publication of *On the Origin of Species*
 iii. Voyages on the HMS *Beagle*
 a. i, ii, iii
 b. ii, i, iii
 c. i, iii, ii
 d. iii, ii, i
 e. ii, iii, i

24. Darwin began his academic career at the University of Edinburgh, where he planned to study:
 a. evolution.
 b. genetics.
 c. entomology (insects).
 d. birds.
 e. medicine.

25. What was Darwin's job aboard the HMS *Beagle*?
 a. captain
 b. ship's cook
 c. navigator
 d. naturalist
 e. ship's surgeon

WHAT'S IT ALL ABOUT?

Here's a question to help you pull together what you've learned so far using this text.

Question: Imagine the Earth in 1 million years. We're gone, but our legacy of global warming remains. Regions of the world that were semitropical in the twenty-first century are now tropical, and artic regions are more temperate. Given what you know about genetics, natural selection, and common descent with modification, what is your prediction for life on Earth under these conditions?

What do I do now?
Remember the drill—decide what the question is asking you to do, collect your evidence from this chapter (and the others you've studied), and write!

CHAPTER 17   THE MEANS OF EVOLUTION: MICROEVOLUTION

Basic Chapter Concepts

- Populations evolve because of a change in the frequency of alleles within the gene pool.
- The five agents of microevolution are (1) mutation, (2) migration, (3) genetic drift, (4) nonrandom mating, and (5) natural selection.
- Natural selection acts on individuals within a population, determining the direction of evolution for the population as a unit.

CHAPTER SUMMARY

17.1   What Is It That Evolves?
- There is confusion as to what is the basic unit of evolution: is it a gene, a species, or a population?
- Genes are the raw material for evolution, and populations are the units that evolve.

17.2   Evolution as a Change in the Frequency of Alleles
- Evolution, which is the change in form and function of a population over many generations, is driven by changes in DNA.
- Because a diploid organism has two alleles for any given gene, many allelic variations of a gene can co-exist within a population. These allelic variations translate into small phenotypic variations among members of the population.

17.3   Five Agents of Microevolution
- Mutations, because they change the information encoded in DNA, can drive evolution. However, mutations that add new, adaptive information are extremely rare, so mutation is a minor agent of evolutionary change.
- More commonly, alleles are added to or removed from a population by migration of members into or out of the population. This gene flow can be a major driving force for change if the new members represent the contribution of a different gene pool.
- Small populations may be significantly affected by the random loss of alleles due to disease, natural disaster, or migration into a new habitat. In these cases, alleles are not selected by environmental factors; they are lost without regard to their adaptive value. Such losses cause genetic drift within a population—"drifting" toward a change in allelic frequencies.
- Allelic representation in future generations can change if ancestral organisms carrying that gene mate more frequently than those that don't carry the gene. Nonrandom mating occurs during sexual selection by females for a particular male phenotype; males possessing the desirable phenotype mate more, thus increasing the frequency of the alleles for that phenotype in subsequent generations.
- Natural selection favors the survival of the organisms within a given population that best "fit" in the environment at a given time, thereby forcing populations to adapt to changing environmental conditions.
- Fitness, then, is a relative concept; it can be defined for an individual only relative to other members of the population, and only for a given set of environmental conditions.

17.4   Natural Selection and Evolutionary Fitness
- Environmental pressures select an optimum phenotype for survival, thus increasing the proportion of alleles in succeeding generations that encode for a successful phenotype. Over many generations of natural selection, the frequency of a given allele within a population's gene pool may increase or decrease.

17.5 Three Modes of Natural Selection

- Stabilizing selection acts on polygenic characteristics by favoring "average" forms over the extremes, such as in the case of infant birth weights.
- Directional selection acts to move a population toward expressing one of the extreme forms of a phenotype, as in the case of the moth that becomes darker in response to a sooty environment.
- Disruptive selection drives a character toward both of its extreme forms, a rare occurrence in nature.

WORD ROOTS

macro- = large scale (e.g., *macro*evolution refers to major evolutionary changes)

micro- = small scale (e.g., *micro*evolution refers to evolutionary changes within a species, usually over a short period of time)

KEY TERMS

adaptation _____

allele _____

bottleneck effect _____

directional selection _____

disruptive selection _____

evolution _____

fitness _____

founder effect _____

gene flow _____

gene pool _____

genetic drift _____

genotype _____

macroevolution _____

microevolution _____

migration _____

natural selection _____

phenotype _____

polygenic _____

population _____

sexual selection _____

species _____

stabilizing selection _____

FLASH CARDS

To use the flash cards, tear the page from the book and cut along the dashed lines. The key term appears on one side of the flash card, and its definition appears on the opposite side.

adaptation	founder effect
allele	gene flow
bottleneck effect	gene pool
directional selection	genetic drift
disruptive selection	genotype
evolution	macroevolution
fitness	microevolution

the phenomenon by which an initial gene pool for a population is established by means of that population's migrating to a new area; one of the conditions that potentiates genetic drift

a modification in the form, physical functioning, or behavior of organisms in a population over generations in response to environmental change

the movement of genes from one population to another

one of the alternative forms of a single gene; in pea plants, a single gene codes for seed color, and it comes in two alleles—one codes for yellow seeds, the other for green seeds

the entire collection of alleles in a population

a change in allele frequencies in a population, due to chance, following a sharp reduction in the population's size; one of the factors that potentiates genetic drift

the chance alteration of allele frequencies in a population, with such alterations having greatest impact on small populations

in evolution, the type of natural selection that moves a character toward one of its extremes

the genetic makeup of an organism, including all the genes that lie along its chromosomes

in evolution, the type of natural selection that moves a character toward both of its extremes, operating against individuals that are average for that character; this type of selection seems to be less common in nature than either stabilizing or directional selection

evolution that results in the formation of new species or other large groupings of living things

any genetically based phenotypic change in a population of organisms over successive generations; evolution can also be thought of as the process by which species of living things can undergo modification over successive generations, with such modification sometimes resulting in the formation of new species

a change of allele frequencies in a population over a short period of time; the basis for all large-scale or macroevolution

in evolution, the success of an organism, relative to other members of its population, in passing on its genes to offspring; fitness is a relative concept only; some organisms have more success than others at passing on their genes in a given environment at a given point in time

migration

population

natural selection

sexual selection

phenotype

species

polygenic

stabilizing selection

all the members of a species that live in a defined geographic region at a given time

a regular movement of animals from one location to a distant location; also, the movement of individuals from one population into the territory of another population; migration is the basis of gene flow among populations

a form of natural selection that produces differential reproductive success based on differential success in obtaining mating partners

a process in which the differential adaptation of organisms to their environment selects those traits that will be passed on with greater frequency from one generation to the next

a group of actually or potentially interbreeding natural populations that are reproductively isolated from other such populations

a physical function, bodily characteristic, or action of an organism

in evolution, the type of natural selection in which intermediate forms of a given character are favored over either extreme; this process tends to maintain average traits for a character

having multiple genes affecting a given character, such as height in humans

SELF TEST

Once you have finished studying this chapter, close your books, grab a pencil, and spend the next 15 to 20 minutes completing this practice test.

Compare and Contrast

For each of the following paired terms, write a sentence of comparison ("Both") and a sentence of contrast ("However,").

microevolution/macroevolution
gene flow/genetic drift
gene pool/population
fitness/adaptation
population/species

Short Answer

1. What is the driving force behind microevolution?

2. How can you tell which individuals in a population have the greatest relative fitness?

3. How can sexual selection affect the frequency of alleles within the gene pool?

4. Why isn't inbreeding, or mating among relatives, considered an evolutionary mechanism?

5. A population of field mice lived in an area that was isolated from other populations by a rather deep lake. However, developers soon discovered that the lake would be a great place to build vacation homes. Speculate on what might happen to the field mouse population in terms of natural selection and the various changes that could occur.

Multiple Choice

Circle the letter that best answers the question.

1. Which of the following can contribute to genetic drift?
   a. small population size
   b. a population bottleneck
   c. the founder effect
   d. environmental disasters
   e. all of the above

2. Assortative mating is based on:
   a. allele frequency.
   b. phenotype.
   c. the founder effect.
   d. randomness of mate selection.
   e. none of the above

3. Back in the fossil field, you have discovered many fossils that are similar to *Neko borgus*. You notice however, that the more recent *Neko* fossils tend to be larger than those that are more ancient. From this you conclude that:
   a. *Neko* has been subjected to directional selection.
   b. *Neko* species continue to grow after death.
   c. *Neko* has been subjected to stabilizing selection.
   d. Environmental conditions have caused the *Neko* fossils to expand over time.
   e. *Neko* has been subjected to disruptive selection.

4. Evidence for evolution is found in the:
   a. genes of an individual.
   b. phenotypes of individual organisms.
   c. allele frequencies in a population.
   d. birds with small beaks.
   e. birds with large beaks.

5. What is the consequence of genetic drift?
   a. improvement of the population
   b. increase in strength of the population
   c. decrease in IQ of the population
   d. unpredictable effects on characters in a population
   e. increase in IQ of the population

6. Which of the following processes is adaptive?
   a. genetic drift
   b. mutation
   c. gene flow
   d. natural selection
   e. all of the above

7. The emergence of very diverse species of house pets (birds, domestic dogs) is an example of:
   a. primary evolution.
   b. secondary evolution.
   c. macroevolution.
   d. microevolution.
   e. minievolution.

8. Of the following, which consistently pushes populations toward a better "fit" with their environment?
   a. nonrandom mating
   b. genetic drift
   c. natural selection
   d. gene flow
   e. founder effect

9. The relative fitness of different types within a population is probably:
   a. unchanging.
   b. dependent on temperature.
   c. changeable.
   d. dependent on size of the population
   e. all of the above

10. The number of individuals within a population with a certain genotype is predicted by:
    a. Mendel's Law of Segregation.
    b. Darwin's scale of biodiversity.
    c. the Hardy-Weinberg theorem.
    d. the Pythagorean theorem.
    e. Mendel's Law of Independent Assortment.

11. What prevents a population from becoming genetically isolated?
    a. genetic drift
    b. the size of the population
    c. bottleneck event
    d. random mating
    e. sexual selection

Match the following terms with their description. Each choice may be used once, more than once, or not at all.
    a. genetic drift
    b. directional selection
    c. gene pool
    d. microevolution
    e. fitness

12. _____ Favors the extreme range of a phenotype

13. _____ Change in gene frequency in a small population

14. _____ The total alleles existing in a population

15. _____ Relative measure of the ability to reproduce

16. _____ Genetic change in populations causing new phenotypes to appear

17. From the following list of molecules, where would a genetic mutation first appear?
    a. DNA
    b. protein
    c. RNA
    d. enzyme
    e. carbohydrate

18. The evolutionary force that operates on the basis of random variation is:
    a. natural selection.
    b. migration.
    c. genetic drift.
    d. sexual selection.
    e. isolation.

19. The work of Peter and Rosemary Grant in the Galápagos deals with:
    a. duplicating the work of Charles Darwin.
    b. demonstrating natural selection in action.
    c. showing that global warming is affecting the Darwin finches.

    d. illustrating the Hardy-Weinberg theorem.
    e. showing that evolution occurs only over large periods of time.

20. The often-used "bell curve" that usually results when test scores are compared with a number of students is an example of what type of selection?
    a. disruptive
    b. stabilizing
    c. directional
    d. convergent
    e. variable

21. A species is composed of:
    a. organisms that live together.
    b. a group of reproductive females.
    c. populations that have the potential to interbreed and produce fertile offspring.
    d. organisms located in the same habitat.
    e. all males and females in the same population.

22. Caterpillars that can maintain coloration similar to leaves avoid predation by birds. What kind of selection is this?
    a. adaptive
    b. stabilizing
    c. directional
    d. sexual
    e. disruptive

23. Male peacocks have large tails and are very colorful and showy, whereas female peacocks are rather dull. This is an example of:
    a. genetic drift.
    b. macroevolution.
    c. founder effect.
    d. sexual selection.
    e. fitness.

24. Which of the following is a source of new genetic combinations in a population?
    a. random mating
    b. mitosis
    c. mutation
    d. genetic drift
    e. gene flow

25. The cocker spaniel has problems that have arisen from centuries of selective breeding. This is an example of:
    a. too many new genes in a population.
    b. what random breeding can do to a population.
    c. a loss of genetic variation in a population.
    d. sexual selection.
    e. microevolution.

WHAT'S IT ALL ABOUT?

Here's a question to help you pull together what you've learned so far using this text.

Question: Scientists have puzzled about the evolutionary processes needed to generate complex structures, such as eyes. Could a rudimentary eye (a patch of cells that detect light) have developed slowly by accumulating mutations, or suddenly, by many mutations occurring all at once? Given what you know about mutations and natural selection, do you think complex structures develop slowly or suddenly?

What do I do now?

Remember the drill—decide what the question is asking you to do, collect your evidence from this chapter (and the others you've studied), and write!

CHAPTER 18   THE OUTCOMES OF EVOLUTION: MACROEVOLUTION

Basic Chapter Concepts

- The basic unit of life forms is the species—a natural population of interbreeding organisms that are reproductively isolated from other populations of organisms that they could or do interbreed with.
- New species occur when populations cease to interbreed through the development of some reproductively isolating mechanism.
- Each species can be identified by a unique binomial name and placed in a taxonomy according to the degree of relatedness to other species.

CHAPTER SUMMARY

18.1   What Is a Species?
- Species are groups of natural populations that are reproductively isolated from other species. Thus, breeding behavior is the defining feature of a species.

18.2   How Do New Species Arise?
- New species arise when populations of interbreeding organisms reproduce after being isolated from each other. The parental species may be transformed over time into a dramatically different species through the action of natural selection (anagenesis), or a new species may branch off from the parental species as a result of some force that reproductively isolates the offspring from the parents (cladogenesis).
- Geographic separation is necessary to achieve speciation, but it must be accompanied by some other isolating mechanism to ensure that the separated populations accumulate enough phenotypic changes while separated to make interbreeding impossible should the two populations be reunited. Geographic separation is an extrinsic isolating mechanism.
- Six intrinsic reproductive isolating mechanisms foster the process of speciation in the absence of geographic isolation. Closely related organisms that occupy different habitats, do not mate in the same season, or do not share courtship rituals will remain separate species despite geographic proximity. Physical differences between the two related species—such as disparities in genital organs that prevent mating, genetic incompatibilities between gametes, or infertility of the resulting zygote—can encourage speciation.

18.3   Many New Species from One: Adaptive Radiation
- Speciation is more likely to occur in a species that has a specialized lifestyle in a given environment. Likewise, environments that provide many resources and habitats to exploit encourage speciation.

18.4   The Pace of Speciation
- The fossil record suggests that speciation occurs rapidly, with long periods of stability between speciation events.

18.5   The Categorization of Earth's Living Things
- Taxonomic classification groups species into categories according to the degree of relatedness. Related species share a common ancestor, and the degree of relatedness depends on how long ago that ancestor lived.
- Classification requires that each species have a specific and universally understood name. The binomial (two-name) naming system was developed by Carl von Linné (Linnaeus, in Latin) in the eighteenth century.

18.6 Classical Taxonomy and Cladistics

- Classical taxonomy uses physical structure (morphology) to fit animals within the evolutionary tree, comparing modern forms with fossil forms to determine relatedness. DNA sequence information also provides an additional basis for judging relatedness. Classical taxonomy seeks both to group species that share similar features and to describe evolutionary relationships.
- Cladistics uses characteristics present in a common ancestor and derived characteristics unique to a given taxon to determine the relatedness. Closely related species are assumed to have more derived characteristics in common. Cladistics, then, is concerned with establishing a line of descent.

WORD ROOTS

allo- = different (e.g., *allo*steric regulation of enzymes takes place at a site different from the active site)

clad- = branch (e.g., *clad*istics groups organisms into separate classifications according to their shared characteristics)

taxon = group (e.g., comparative morphologists might study limb structure among *taxa,* or groups of organisms)

KEY TERMS

adaptive radiation ⎯⎯⎯⎯⎯⎯⎯⎯⎯⎯

allopatric speciation ⎯⎯⎯⎯⎯⎯⎯⎯⎯

analogy ⎯⎯⎯⎯⎯⎯⎯⎯⎯⎯⎯⎯⎯⎯

ancestral character ⎯⎯⎯⎯⎯⎯⎯⎯⎯⎯

binomial nomenclature ⎯⎯⎯⎯⎯⎯⎯⎯

biological species concept ⎯⎯⎯⎯⎯⎯⎯

cladistics ⎯⎯⎯⎯⎯⎯⎯⎯⎯⎯⎯⎯⎯

cladogram ⎯⎯⎯⎯⎯⎯⎯⎯⎯⎯⎯⎯

class ⎯⎯⎯⎯⎯⎯⎯⎯⎯⎯⎯⎯⎯⎯⎯

convergent evolution ⎯⎯⎯⎯⎯⎯⎯⎯⎯

derived character ⎯⎯⎯⎯⎯⎯⎯⎯⎯⎯

domain ⎯⎯⎯⎯⎯⎯⎯⎯⎯⎯⎯⎯⎯⎯

extrinsic isolating mechanism ⎯⎯⎯⎯⎯

family ⎯⎯⎯⎯⎯⎯⎯⎯⎯⎯⎯⎯⎯⎯

genus ⎯⎯⎯⎯⎯⎯⎯⎯⎯⎯⎯⎯⎯⎯⎯

homology ⎯⎯⎯⎯⎯⎯⎯⎯⎯⎯⎯⎯⎯

intrinsic isolating mechanism ⎯⎯⎯⎯⎯

kingdom ⎯⎯⎯⎯⎯⎯⎯⎯⎯⎯⎯⎯⎯⎯

order ⎯⎯⎯⎯⎯⎯⎯⎯⎯⎯⎯⎯⎯⎯⎯

phylogeny ⎯⎯⎯⎯⎯⎯⎯⎯⎯⎯⎯⎯⎯

phylum ⎯⎯⎯⎯⎯⎯⎯⎯⎯⎯⎯⎯⎯⎯

polyploidy ⎯⎯⎯⎯⎯⎯⎯⎯⎯⎯⎯⎯⎯

reproductive isolating mechanisms ⎯⎯⎯

speciation ⎯⎯⎯⎯⎯⎯⎯⎯⎯⎯⎯⎯⎯

species ⎯⎯⎯⎯⎯⎯⎯⎯⎯⎯⎯⎯⎯⎯

sympatric speciation ⎯⎯⎯⎯⎯⎯⎯⎯⎯

systematics ⎯⎯⎯⎯⎯⎯⎯⎯⎯⎯⎯⎯

taxon ⎯⎯⎯⎯⎯⎯⎯⎯⎯⎯⎯⎯⎯⎯⎯

FLASH CARDS

To use the flash cards, tear the page from the book and cut along the dashed lines. The key term appears on one side of the flash card, and its definition appears on the opposite side.

adaptive radiation	cladogram
allopatric speciation	class
analogy	convergent evolution
ancestral character	derived character
binomial nomenclature	domain
biological species concept	extrinsic isolating mechanism
cladistics	family

an evolutionary tree constructed using the cladistic system	the rapid emergence of many species from a single species that has been introduced to a new environment; the different species specialize to fill available niches in the new environment
a taxonomic grouping subordinate to phylum and superordinate to order; humans are in the class Mammalia	speciation that involves the geographic separation of populations; most speciation involves geographic separation, followed by the development of intrinsic isolating mechanisms in the separated populations
evolution that occurs when similar environmental influences shape two separate evolutionary lines in similar ways	a structure found in different organisms that is similar in function and appearance but is not the result of shared ancestry; analogies must be distinguished from homologies to get a true picture of evolutionary relationships
a character unique to groupings of organisms (taxa) descended from a common ancestor	a character that existed in the common ancestor of a group of organisms; cladistics distinguishes ancestral from derived characters and uses these characters to determine evolutionary relationships
the highest-level taxonomic grouping of organisms; there are only three domains: Archaea, Bacteria, and Eukarya	the system of naming species that uses two names (genus and species) for each species; this system helps identify groupings among living things
a barrier to interbreeding of populations that is not an inherent characteristic of the organisms in the populations; geographic barriers, such as rivers, are extrinsic isolating mechanisms	a definition of species that relies on the breeding behavior of populations in nature; it defines *species* as groups of actually or potentially interbreeding natural populations that are reproductively isolated from other such groups
a taxonomic grouping subordinate to order and superordinate to genus	the branch of systematics that uses shared derived characters to determine the order of branching events in speciation and therefore which species are most closely related; cladistics is concerned only with evolutionary relationships, not classification

genus

polyploidy

homology

reproductive isolating mechanisms

intrinsic isolating mechanism

speciation

kingdom

species

order

sympatric speciation

phylogeny

systematics

phylum

taxon

a form of sympatric speciation in which one or more sets of chromosomes are added to the genome of an organism; human beings cannot survive in a polyploid state, but many plants flourish in it; polyploidy is a means by which speciation can occur (most often in plants) in a single generation

a taxonomic grouping of related species; this category is subordinate to family and superordinate to species; humans are in the genus *Homo*

any factor that, in nature, prevents interbreeding between individuals of the same or closely related species

a structure that is shared in different organisms owing to inheritance from a common ancestor; homologies are used to help decipher evolutionary relationships

the development of new species through evolution

a difference in anatomy, physiology, or behavior that prevents interbreeding between individuals of the same species or of closely related species; one or more intrinsic isolating mechanisms must develop for two populations of the same species to evolve into separate species

a group of actually or potentially interbreeding natural populations that are reproductively isolated from other such populations

a taxonomic grouping superordinate to every other grouping except domain; there are four kingdoms in Domain Eukarya: Protista, Fungi, Animalia, and Plantae

a type of speciation that occurs in the absence of the geographic separation of populations; polyploidy is a special form of sympatric speciation

a taxonomic grouping subordinate to class and superordinate to family; humans are in the Order Primates

the field of biology dealing with the diversity and relatedness of organisms; systematists study the evolutionary history of groups of organisms

a hypothesis about the evolutionary relationships of a group of organisms

any of the categories used in the classification of Earth's organisms, both living and extinct; in order of increasing inclusiveness, these categories are species, genus, family, order, class, phylum, kingdom, and domain

a category of living things, directly subordinate to the category of kingdom, whose members share traits as a result of shared ancestry

SELF TEST

After you have finished studying this chapter, close your books, grab a pencil, and spend the next 15 to 20 minutes completing this practice test.

Compare and Contrast

For each of the following paired terms, write a sentence of comparison ("Both") and a sentence of contrast ("However,").

cladogenesis/anagenesis
extrinsic isolating mechanism/intrinsic isolating
   mechanism
phylum/kingdom
analogous/homologous
systematics/cladistics

Short Answer

1. What is the difference between sympatric and allopatric speciation?
2. Describe the process of adaptive radiation.
3. What is the major difference between the gradualism and punctuated equilibrium models?
4. In cladistic analysis, what determines the degree of relatedness among species?
5. What is a species?
6. How do new species arise?

Multiple Choice

Circle the letter that best answers the question.

1. The biological species concept fails to recognize species of:
   a. primates.
   b. bacteria.
   c. vertebrates.
   d. trees.
   e. fish.

2. Two organisms of different species mate and produce offspring. This offspring cannot mate with its siblings, nor with members of either parental species. This is an example of:
   a. hybrid infertility.
   b. gametic isolation.
   c. behavioral isolation.
   d. ecological isolation.
   e. temporal isolation.

3. In the Galapagos finches, adaptive radiation occurred _____ islands, whereas allopatric speciation occurred _____ islands.
   a. within; on large
   b. between; within
   c. on large; on small

   d. within; between
   e. on small; on large

4. The most inclusive category is:
   a. kingdom.
   b. genus.
   c. phylum.
   d. domain.
   e. order.

5. Convergent evolution can result in species with:
   a. homologous structures.
   b. systematic structures.
   c. analogous structures.
   d. derived characters.
   e. none of the above

6. Which of the following is used to determine whether two populations belong to the same species?
   a. ability to interbreed
   b. color
   c. size range
   d. habitat choice
   e. diet

7. Two individuals from different populations attempt to mate but are unable to successfully coordinate the mating dance. This is an example of a(n) _____ isolating mechanism.
   a. ecological
   b. temporal
   c. behavioral
   d. mechanical
   e. gametic

8. Speciation as a small series of changes that accumulate over time is known as:
   a. punctuated equilibrium.
   b. cladistics.
   c. gradualism.
   d. stasis.
   e. compounded modification.

9. Classification of species based on their relatedness is known as:
   a. eurythmics.
   b. systematics.
   c. rhetorics.
   d. mathematics.
   e. aeronautics.

10. You are doing a cladistic analysis on a group of five species. How many would you expect to share the most ancient derived trait?
   a. 1
   b. 2
   c. 3
   d. 4
   e. 5

11. You now begin the task of grouping your many *Neko* fossils. Can you determine which examples are from the same species by using the biological species concept?
    a. No, because the specimens are sympatric.
    b. No, because the specimens are allopatric.
    c. No, because you cannot determine the potential for interbreeding between samples.
    d. Yes, because of the similarities of the skeleton to other modern species.
    e. Yes, because as long as the samples are biological, they will satisfy the concept definition.

12. Plants are much more likely than animals to form hybrids. Most of these hybrids are sterile, but occasionally these individuals can form the basis for a new species. The reason is that plants, unlike animals, are capable of surviving:
    a. polyploidy.
    b. aploidy.
    c. punctuated equilibrium.
    d. convergent evolution.
    e. harsh winters.

13. The term *systematics* is best defined as:
    a. studying speciation that takes place in brief bursts, separated by periods of stasis.
    b. studying speciation that occurs as a result of geographic separation.
    c. studying speciation that occurs in the absence of geographic separation.
    d. studying evolutionary history of groups of organisms.
    e. studying several new species that have descended from one species and become more specialized to specific habitats.

14. In punctuated equilibrium:
    a. speciation takes place in brief bursts, separated by periods of stasis.
    b. speciation occurs due to geographic separation.
    c. speciation occurs in the absence of geographic separation.
    d. speciation is studied only in asexual species
    e. several new species descend from one species and become more specialized to specific habitats.

15. The term *sympatric speciation* is best defined as:
    a. speciation that takes place in brief bursts, separated by periods of stasis.
    b. speciation that occurs as a result of geographic separation.
    c. speciation that occurs in the absence of geographic separation.

    d. speciation that occurs as a result of historical separation.
    e. several new species descending from one species and becoming more specialized to specific habitats.

16. In allopatric speciation:
    a. speciation occurs as a result of historical separation.
    b. speciation takes place in brief bursts, separated by periods of stasis.
    c. speciation occurs as a result of geographic separation.
    d. speciation occurs in the absence of geographic separation.
    e. several new species descend from one species and become more specialized to specific habitats.

17. In adaptive radiation:
    a. speciation takes place in brief bursts, separated by periods of stasis.
    b. speciation occurs as a result of geographic separation.
    c. speciation occurs as a result of historical separation.
    d. speciation occurs in the absence of geographic separation.
    e. several new species descend from one species and become more specialized to specific habitats.

18. Which of the following are *not* intrinsic isolating mechanisms?
    a. ecological
    b. behavioral
    c. temporal
    d. spatial
    e. gametic

19. The correct order, from least to most inclusive, of the classification categories is:
    a. species, genus, domain, order, class, phylum, kingdom, family.
    b. species, genus, class, family, order, phylum, kingdom, domain.
    c. species, genus, family, order, class, kingdom, phylum, domain.
    d. species, genus, family, class, order, phylum, kingdom, domain.
    e. species, genus, family, order, class, phylum, kingdom, domain.

20. On official documents, you are asked to give your first and last names. Which name is analogous to a genus, and which to a species name?
    a. Both first and last names are analogous to genus.

b. Both first and last names are analogous to species.

c. First name is analogous to species, last name to genus.

d. First name is analogous to genus, last name to species.

e. There is no analogous structure between them.

21. Suppose that two populations of bunnies are separated from one another by the change in the course of a stream. After many generations, a rock bridge forms across the stream, and the two groups can now interbreed. However, no hybrid offspring appear. This indicates that _____ may be isolating them.

a. temporal mechanisms

b. ecological mechanisms

c. geological mechanisms

d. hybrid inviability

e. sympatric speciation

22. The sequences of the rRNA genes from bacterial isolates A, B, and C were determined. A and B differ at 20 nucleotides; A and C differ at 45 nucleotides; and B and C differ at 5 nucleotides. None of these isolates mates with one another. Are any of these isolates likely to be the same bacterial species?

a. A and B are likely to be the same species.

b. A and C are likely to be the same species.

c. B and C are likely to be the same species.

d. A, B and C are likely to be the same species.

e. It is unlikely that any of these belong to the same species.

Use the chart below to answer Questions 23, 24, and 25. Imagine that the following are all species of small mammals recently discovered in the Amazonian forests. They all bear an overall similarity in body shape. Your job is to build a phylogenetic tree that best represents their evolutionary history.

Character/ Species	A	M	L	B	R
Six toes	X	X			X
Blue eyes	X			X	X
Whiskers				X	
Meat-eating	X	X	X	X	X
Very small size		X			X

23. Which character is likely to be ancestral?

a. six toes

b. blue eyes

c. whiskers

d. meat-eating

e. small size

24. Which is likely to be the most recently derived character?

a. six toes

b. blue eyes

c. whiskers

d. meat-eating

e. large size

25. Which two species branched most recently (i.e., share the greatest number of characteristics)?

a. A and L

b. A and B

c. A and R

d. M and L

e. M and B

WHAT'S IT ALL ABOUT?

Here's a question to help you pull together what you've learned so far using this text.

Question: If we look at the phylogenetic trees within this chapter in the textbook, we see a trend from comparatively simple organisms on the leftmost branches to more complex organisms on the right branches. Why does the process of evolution lead to increasing complexity?

What do I do now?
Remember the drill—decide what the question is asking you to do, collect your evidence from this chapter (and the others you've studied), and write!

CHAPTER 19   A SLOW UNFOLDING: THE HISTORY OF LIFE ON EARTH

Basic Chapter Concepts

- The geological time scale begins 4.6 billion years ago with Earth's formation and is divided into time periods determined by the type of fossils that have accumulated.
- Life, defined as self-replicating molecules, appeared on Earth from 3.5 to 4 billion years ago. All life can be traced to a universal ancestor.
- The Cambrian Explosion marks a remarkable increase in the diversity of animal forms and an increase in the rate of evolution of new animal forms.
- The human family tree begins as chimpanzees and hominids diverge from a common primate ancestor. The most commonly accepted hypothesis, the "out-of-Africa" hypothesis, proposes that modern humans evolved in Africa and then migrated over the globe.

CHAPTER SUMMARY

19.1 The Geological Timescale: Life Marks Earth's Ages
- The geological time scale is divided into eras, which are subdivided into periods, which are divided into epochs. The boundaries between the time periods are marked by major changes in the fossil record.

19.2 How Did Life Begin?
- "Life" probably began around deep-sea vents or hot-spring pools as organic molecules reacted with each other to form more complex molecules.
- Many scientists believe that RNA molecules were the first self-replicating molecules, because RNA can serve as a template for the transfer of information as well as a catalyst for chemical reactions.

19.3 The Tree of Life
- All living things on Earth descended from a "universal ancestor." Three branches sprout from this ancestor—the Domains Bacteria, Archaea, and Eukarya.

19.4 A Long First Era: The Precambrian
- The Precambrian era, which began with Earth's formation, is defined by the appearance of life in the form of bacteria. These bacteria are responsible for producing oxygen through the reactions of photosynthesis.
- Animals first appeared in the Precambrian era, about 600 Mya. The mass extinctions that accompanied the "oxygen holocaust" opened up niches that made possible the diversification of life forms, in turn making the Cambrian Explosion possible.

19.5 The Cambrian Explosion
- Animal-like multicellular organisms first appeared about 575 Mya, accounting for 35 of the 36 animal phyla known today.
- Animals of the Cambrian period were large with bizarre shapes, and many of them thrived on the floor of the ancient seas. Scientists have speculated that the abundance of oxygen in the atmosphere may have fueled this increase in size.

19.6 The Movement onto the Land: Plants First
- Plants were the first organisms to make the transition from the seas to the land, about 460 Mya. These plants adapted to the drier land environment by developing a waxy cuticle to minimize water loss and a supportive vascular system to counter the effect of gravity.
- The major adaptation, however, was shifting the maturation of the embryos to a site within the parent instead of outside in the water.

19.7   Animals Follow Plants onto the Land

- Animals followed plants onto land about 400 Mya; the first were wingless insects, followed by the lobe-finned fishes, which are the ancestors of all tetrapod vertebrates. Another critical adaptation to live on land was the appearance of the amniotic egg—a structure that surrounds the developing embryo with padding, food, and a hard, protective covering.
- Mammals appeared about 220 Mya, during the time of the dinosaurs. Mammalian evolution did not take off until after the sudden extinction of the dinosaurs opened niches, about 100 Mya.
- Primates, the ancestors of the modern chimpanzees and humans, appeared 60 Mya. Tracing the evolution of modern humans depends on interpreting structural and DNA clues present in fossil skeletons.
- Because evidence (both fossil and DNA) of human evolution is incomplete, several possible family trees have been proposed. Most scientists agree that humans evolved from African hominids, abandoning a tree-dwelling lifestyle and adopting an upright stance and bipedalism.

WORD ROOTS

angio- = vessel; **sperm** = seed (e.g., an *angiosperm* is a plant that has flowers, and its egg-containing ovules mature into seeds enclosed by tissues)

archaea = primitive (e.g., *archaea* are a group of primitive prokaryotic species)

euk- = easily formed; **-karya** = kernel (e.g, *eukarya* are a group of species that are composed of single cells or groups of cells that have a nucleus)

gymno- = naked; **sperm** = seed (e.g., a *gymnosperm* is a plant with seeds that are not enclosed in an ovary; the conifers are an example)

ribo- = ribose; **-zyme** = enzyme (e.g., a *ribozyme* is an RNA molelcule that can behave like an enzyme)

KEY TERMS

amniotic egg _____

angiosperm _____

Archaea _____

Bacteria _____

bryophytes _____

Cambrian Explosion _____

continental drift _____

Cretaceous Extinction _____

Eukarya _____

eukaryote _____

gymnosperm _____

Permian Extinction _____

phylum _____

prokaryote _____

ribozyme _____

seedless vascular plant _____

FLASH CARDS

To use the flash cards, tear the page from the book and cut along the dotted lines. The key term appears on one side of the flash card, and its definition appears on the opposite side.

amniotic egg	bryophytes
angiosperm	Cambrian Explosion
Archaea	continental drift
Bacteria	Cretaceous Extinction

plants that lack a vascular (fluid transport) structure; bryophytes are the most primitive of the four principal varieties of plants; mosses are the most familiar example

an egg with a hard outer casing and an inner series of membranes and fluids that provide protection, nutrients, and waste disposal for a growing embryo; the evolution of the amniotic egg, in reptiles, freed them from the constraint of having to reproduce near water

an alleged sudden evolution of the ancestors of modern animals beginning about 542 million years ago

a flowering seed plant whose seeds are enclosed within the tissue called fruit; the most dominant and diverse of the four principal types of plants; examples include roses, cacti, corn, and deciduous trees

the lateral movement of continental plates over the globe, allowing continents to divide and rejoin in different patterns; this process can separate populations of organisms, providing the geographic barriers that can result in speciation

with Bacteria and Eukarya, one of three domains of the living world, composed solely of microscopic, single-celled organisms superficially similar to bacteria but genetically quite different; many live in extreme environments, such as boiling-hot vents on the ocean floor

a mass extinction event that occurred at the boundary between the Cretaceous and Tertiary periods; this event, which included an asteroid impact, resulted in the extinction of the dinosaurs along with many other organisms

with Archaea and Eukarya, one of three domains of the living world, composed solely of single-celled, microscopic organisms that superficially resemble Archaea but are genetically quite different

Eukarya

phylum

eukaryote

prokaryote

gymnosperm

ribozyme

Permian Extinction

seedless vascular plant

a category of living things, directly subordinate to the category of kingdom, whose members share traits as a result of shared ancestry

with Bacteria and Archaea, one of three domains of the living world, composed of four subordinate kingdoms: Plantae, Animalia, Fungi, and Protista; all members are composed of single cells or groups of cells that have a nucleus

a single-celled organism whose complement of DNA is not contained within a nucleus

an organism that is a member of Domain Eukarya

molecule composed of RNA that can both encode genetic information and act as an enzyme; molecules similar to these would have been necessary early in the evolution of life

a seed plant whose seeds do not develop within fruit tissue; one of the four principal varieties of plants, gymnosperms reproduce through wind-aided pollination; coniferous trees, such as pine and fir, are the most familiar examples

a plant that has a vascular (fluid transport) structure but that does not reproduce through use of seeds; one of the four principal varieties of plants; ferns are the most familiar example

the greatest mass extinction event in Earth's history; this event, in which up to 96 percent of all species on Earth were wiped out, occurred about 245 million years ago

SELF TEST

Once you have finished studying this chapter, close your books, grab a pencil, and spend the next 15 to 20 minutes completing this practice test.

Compare and Contrast

For each of the following paired terms, write a sentence of comparison ("Both") and a sentence of contrast ("However,").

angiosperm/gymnosperm
Archaea/Bacteria
epoch/era
domain/phylum
bryophytes/seedless vascular plants

Short Answer

1. What is the Malnourished Earth hypothesis, and how might it have affected the rate of evolution?

2. What two adaptations allowed plants to colonize land?

3. What is the relationship between the terms *continental drift* and *adaptive radiation*?

4. What is thought to be responsible for the Cretaceous Extinction?

5. How was the first oxygen produced on earth? What are some of the benefits of oxygen production?

6. Of the three domains of life, which are most closely related?

7. When did the Cambrian Explosion occur? What was its significance?

Multiple Choice

Circle the letter that best answers the question.

1. Based on structural similarities, you conclude that your fossil discovery, *Neko borgus*, was a type of early mammal. *Neko borgus* would therefore have been preceded onto land by all of the following organisms *except:*
   a. reptiles.
   b. fish.
   c. plants.
   d. humans.
   e. all of the above

2. Amphibians:
   a. were the first animals to emerge on land.
   b. lay amniotic eggs.
   c. went mostly extinct during the Cambrian Explosion.
   d. live on both land and water.
   e. are members of the Domain Archaea.

3. Which of the following is *not* a special adaptation of plants to land?
   a. waxy cuticle
   b. vascular system
   c. photosynthesis
   d. seeds
   e. flowers

4. Human beings:
   a. are mammals.
   b. evolved from the primate line.
   c. evolved in Africa.
   d. descended from tree-dwelling ancestors.
   e. all of the above

5. What advantage did the hard shells of arthropods provide during the move onto land?
   a. swimming ability
   b. temperature regulation
   c. prevention of desiccation (drying)
   d. increased enzymatic rate
   e. a lighter body, enabling the animal to move faster

6. The domains of life are:
   a. Prokarya, Eukarya.
   b. Bacteria, Eukarya, Archaea.
   c. Plantae, Animalia, Bacteria.
   d. Bacteria, Eukarya.
   e. Protista, Plantae.

7. Photosynthetic organisms produce _____, which protects terrestrial organisms from _____.
   a. oxygen; UV irradiation
   b. oxygen; toxins
   c. UV; irradiation
   d. mitochondria; toxins
   e. water loss; dehydration

8. Which of the following is the most successful plant group (in number of species) that exists today?
   a. bryophytes
   b. ferns
   c. gymnosperms
   d. angiosperms
   e. algae

9. Based on structural similarities, you conclude that your fossil discovery, *Neko borgus*, was a type of early mammal. What domain would you place this species in?
   a. Archaea
   b. Bacteria
   c. Eukarya
   d. Plantae
   e. Protista

10. The "missing link," in the understanding of molecular evolution on early Earth, was the lack of an assembly mechanism for the information storage molecules, DNA and RNA. This link was found with the discovery of the enzyme ribozyme, which is actually a type of:
    a. protein.
    b. carbohydrate.
    c. DNA.
    d. RNA.
    e. ATP.

Match the following terms with their description. Each choice may be used once, more than once, or not at all.
    a. lemur
    b. accretion
    c. Archaea
    d. bryophyte
    e. Bacteria

11. _____ Modern-day descendant of early primates

12. _____ Mosses (seedless, nonvascular plants)

13. _____ Earliest organisms

14. _____ Clumping of cosmic particles into a larger object

15. _____ A prokaryotic domain of life

16. The ancestors of early primate mammals shared which characteristic?
    a. large, front-facing eyes
    b. amniotic eggs
    c. binocular vision
    d. opposable digits
    e. tree-dwelling existence

17. In experiments in which energy is supplied to a sealed chamber containing a mixture of gases simulating the early Earth's atmosphere, what will be formed?
    a. amino acids
    b. sugars
    c. nucleic acids
    d. lipids
    e. starch

18. Notable features of the geological time scale include:
    a. the radiometric dating of heavy metals.
    b. periods of mass extinction.
    c. environmental temperatures.
    d. DNA analysis of the organisms of particular periods.
    e. living fossil organisms.

19. The great evolutionary explosion of animals with fantastic shapes occurred during which geological period?
    a. Cambrian
    b. Precambrian
    c. Devoian
    d. Cretaceous
    e. Carboniferous

20. Pangaea:
    a. was a landmass that eventually broke up to form present day continents.
    b. is the idea that life came from outer space.
    c. is the idea that all life on Earth is related.
    d. is the theory that giant plates of the earth are constantly moving.
    e. refers to a little known theory of evolution.

21. You have a great interest in seeing a living dinosaur. If you had access to a time machine, what era would you set the dial for?
    a. Precambrian
    b. Cenozoic
    c. Paleozoic
    d. Mesozoic
    e. Jurassic

22. The theory of continental drift helps scientists to explain all of the following *except:*
    a. locations of volcanoes.
    b. distribution of plants and animals.
    c. formation of mountain ranges.
    d. earthquakes.
    e. formation of lakes.

23. Which of the following plants are the most complex and evolved last?
    a. gymnosperms
    b. angiosperms
    c. fungi
    d. mosses
    e. spore-bearing plants

24. The greatest extinction the world has ever known occurred at the end of which period?
    a. Cretaceous
    b. Permian
    c. Triassic
    d. Jurassic
    e. Tertiary

25. Which of the following is *not* a component of the atmosphere that Miller used in his experiment?
    a. hydrogen
    b. methane
    c. ammonia
    d. oxygen
    e. water

WHAT'S IT ALL ABOUT?

Here's a question to help you pull together what you've learned so far using this text.

Question: The time scale for life on Earth (Figure 19.3 in the text) shows that life on Earth was rather slow to get started but rapidly diversified, beginning 544 Mya during the Cambrian Explosion. One can argue that evolution can be easily compared to forming a self-replicating living organism. What challenges must be conquered to create that first living organism?

What do I do now?

Remember the drill—decide what the question is asking you to do, collect your evidence from this chapter (and the others you've studied), and write!

CHAPTER 20 ARRIVING LATE, TRAVELING FAR: THE EVOLUTION OF HUMAN BEINGS

Basic Chapter Concepts

- The taxonomic group of human-like primates is known as the Hominini.
- The origin of the homini has been traced back to Africa.
- While there are several theories about the emigration of homini from Africa to Europe and Asia, it is generally agreed that modern humans left Africa less than 50,000 years ago.
- The discovery of the "Hobbit People" fossils has led to competing theories about their relatedness to other recent homini species.

CHAPTER SUMMARY

20.1 The Human Family Tree
- Six to seven million years ago, evolution gave rise to the lineages that became chimpanzees and modern humans. The taxonomic group of human-like primates is known as the Hominini.

20.2 Human Evolution in Overview
- The only surviving species of Hominini is *Homo sapiens*. Until recently, it was thought that most of human evolution took place in east Africa. However, recent discoveries in Chad (the Toumaï fossil, *Sahelanthropus tchadensis*) have led scientists to expand the range westward.

20.3 Interpreting the Fossil Evidence
- Both fossil evidence and DNA analysis are important for determining lineages. However, fossils are still the primary source of information about extinct species. In the classification of Hominini, two features are definitive: tooth structure and bipedalism.

20.4 Snapshots from the Past: Three Hominins
- *Australopithecus afarensis*. The most famous member of this group is Lucy. Although smaller and most likely partly arboreal, Lucy (and her conspecifics) are similar enough to modern humans to be considered one of our ancestors.
- *Homo ergaster*. As typified by Turkana Boy, this species has adaptations in form and cranial capacity that are more similar to ours. *H. ergaster* is thought to have lived about 1.7 million years ago.
- *Homo neanderthalensis*. Active in Asia and Europe from 130,000 to 28,000 years ago, *H. neanderthalensis* is often used to typify the "caveman," although there is evidence of cultural features such as burial of the dead.

20.5 The Appearance of Modern Human Beings
- There are two hypotheses that attempt to explain the appearance of modern humans and Neanderthals in Europe: the "out-of-Africa" and "multiregional" hypotheses. The "out-of-Africa" hypothesis proposes that by the time humans left Africa, their appearance was similar to the fossils found in Europe. The "multiregional" hypothesis proposes multiple emigrations by populations that subsequently interbred to produce *H. sapiens* and *H. neanderthalensis*.
- The earliest fossils of modern humans outside Africa date to 46,000 years ago and were discovered in Australia.
- In Africa, evidence of modern humans from Ethiopia dates back to 195,000 years ago.
- *H. neanderthalensis* became extinct about 12,000 years after the appearance of modern humans in Europe. In Asia, *H. erectus* disappeared about 6,000 years after the appearance of modern humans.

20.6 Next-to-Last Standing? The Hobbit People

- Fossils of *Homo floresiensis*, also known as "Hobbit People," were discovered in Indonesia in 2004.
- *Homo floresiensis* was active until 18,000 years ago.
- There is some controversy about whether *Homo floresiensis* was a separate species or a family group with a medical abnormality (fossil measurements indicate an adult height of about 3 feet).

WORD ROOTS

paleo- = involving or dealing with ancient forms or conditions (e.g., *paleo*biology is concerned with the biology of fossil organisms)

homo- = same; man or human (e.g., *Homo sapiens* are modern human beings)

KEY TERMS

bipedalism _____

Hominini _____

Homo sapiens (H. sapiens) _____

Neanderthal _____

paleoanthropologist _____

FLASH CARDS

To use the flash cards, tear the page from the book and cut along the dotted lines. The key term appears on one side of the flash card, and its definition appears on the opposite side.

bipedalism	Neanderthal
Hominini	**paleoanthropologist**
Homo sapiens (H. sapiens)	

a now-extinct species of hominin that lived in Europe and Asia from approximately 350,000 years ago to 28,000 years ago

among primates, the trait of walking upright, on two legs; along with tooth structure, the most important trait in defining species who are classified as Hominini, meaning members of the taxon of human-like primates

a scientist who studies human evolution through means of analyzing fossils

the taxonomic grouping of human-like primates; modern human beings, or *Homo sapiens,* are one of the species within Hominini; all the other hominin species are now extinct

the species of modern human beings; the last surviving species within the taxonomic category of Hominini, or human-like primates

SELF TEST

Once you have finished studying this chapter, close your books, grab a pencil, and spend the next 15 to 20 minutes completing this practice test.

Compare and Contrast

For each of the following paired terms, write a sentence of comparison ("Both") and a sentence of contrast ("However,").

  Australopithecus/Ardipithecus
  fossil/skeleton
  chimpanzee/human
  arboreal/terrestrial
  tooth structure/molecular biology

Short Answer

1. What does it mean for a species to belong to the Hominini?

2. What is significant about the discovery of *Sahelanthropus tchadensis* in 2002?

3. How is DNA evidence used in paleoanthropology?

4. Who was "Lucy"?

5. Compare and contrast the "out-of-Africa" and "multiregional" hypotheses.

6. What is controversial about the discovery of *Homo floresiensis* in 2004?

Multiple Choice

Circle the letter that best answers the question.

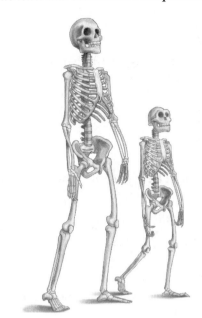

1. Looking at the figure above, which of the following is *not* a discernible difference between "Lucy" and modern humans?
   a. opposable thumbs
   b. smaller stature
   c. longer arms
   d. grasping toes
   e. smaller brain

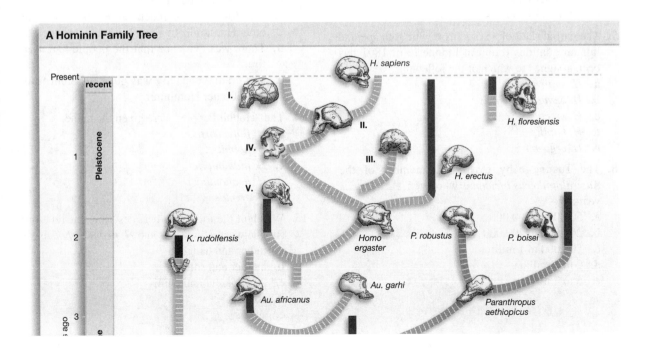

A Hominin Family Tree

2. Looking at the figure on page 189, which of the following species is the immediate ancestor to *H. sapiens*?
   a. I
   b. II
   c. III
   d. IV
   e. V

3. Looking at the figure, which of the following was most recently alive?
   a. I
   b. II
   c. III
   d. IV
   e. V

4. Looking at the figure, which of the following became extinct first?
   a. I
   b. II
   c. III
   d. IV
   e. V

5. Looking at the figure, which of the following was active in Europe?
   a. I
   b. II
   c. III
   d. IV
   e. V

6. Looking at the figure, which pair coexisted?
   a. *H. sapiens* and I
   b. *H. sapiens* and II
   c. *H. sapiens* and III
   d. *H. sapiens* and IV
   e. *H. sapiens* and V

7. The partial skull of an organism with both ape and human qualities found in Indonesia in 1891 has been assigned to which of the following?
   a. *H. sapiens*
   b. *H. neanderthalensis*
   c. *H. erectus*
   d. *H. habilis*
   e. *H. ergaster*

8. The Tuaung baby fossil is a member of the *Australopithecus africanus* who lived _____ years ago.
   a. 50,000 to 100,000
   b. 200,000 to 400,000
   c. 500,000 to 1 million
   d. 2 to 3 million
   e. 6 to 7 million

9. The term used to describe the taxonomic group of human-like primates is:
   a. Apini.
   b. Primatini.
   c. Hominini.
   d. Anthropini.
   e. Parahomini.

10. A human active in Europe 25,000 years ago would be a member of:
    a. *H. sapiens*.
    b. *H. neanderthalensis*.
    c. *H. erectus*.
    d. *H. habilis*.
    e. *H. ergaster*.

11. The "Hobbit People" fossils were found in:
    a. east Africa.
    b. west Africa.
    c. India.
    d. Indonesia.
    e. Mexico.

12. Researchers determined that the cranial capacity of the "Hobbit People" was:
    a. larger than that of a human.
    b. about the same size as that of a human.
    c. about the same size as that of a chimp.
    d. smaller than that of a chimp.
    e. about the same size as that of a dog.

13. The discovery of the "Hobbit People" was stunning for a number of reasons. Which of the following was *not* one of those reasons?
    a. The fossils appeared to be from a more recent time than it was thought other Hominini had survived.
    b. There was evidence of tool use.
    c. The fossils were much larger than those of other Hominini.
    d. There was evidence that the people had used fire.
    e. The cranial capacity was quite different from that of other Hominini.

14. The "Hobbit People" were given the name:
    a. *H. floresiensis*.
    b. *H. hobbit*.
    c. *H. indoamerici*.
    d. *H. neohomini*.
    e. *H. erectus*.

15. Which of the following best describes the relationship between *H. sapiens* and *H. neanderthalensis*?
    a. father and daughter
    b. mother and son
    c. grandparent and child
    d. cousins
    e. sisters

16. The only known fossil of *Homo ergaster* bears similarities to modern humans *except* for which of the following?
    a. height
    b. speech
    c. facial structure
    d. limb length
    e. tool use

17. The single most important defining feature of a hominin is:
    a. bipedalism.
    b. tool use.
    c. mammary glands.
    d. body hair.
    e. opposable thumbs.

18. A realistic model of a specimen can be constructed from fossilized remains by means of:
    a. wind-tunnel modeling.
    b. laser sculpting.
    c. computed tomography.
    d. magnetic resonance imaging.
    e. claymation.

19. The earliest hominin fossils found outside of Africa date to _____.
    a. 20,000
    b. 33,000
    c. 46,000
    d. 59,000
    e. 72,000

20. Approximately _____ of hominin history took place in Africa.
    a. 10 percent
    b. 30 percent
    c. 50 percent
    d. 70 percent
    e. 90 percent

21. Until the 1990s, hominin fossils were found in _____ Africa, then Toumaï fossils were discovered in _____.
    a. western and northern; southern Africa
    b. eastern and southern; western Africa
    c. western and northern; eastern Africa
    d. western and southern; eastern Africa
    e. eastern and northern; southern Africa

22. With the discovery of the Toumaï fossil, the origin of hominins was pushed back to:
    a. 100,000 years ago.
    b. 500,000 years ago.
    c. 1.5 million years ago.
    d. 3 million years ago.
    e. 6 million years ago.

23. The proper name for the Toumaï fossils is:
    a. *Sahelanthropus tchadensis.*
    b. *Australopithecus bahrelghazali.*
    c. *Kenyanthropus platyops.*
    d. *Australopithecus africanus.*
    e. *Ardipithecus kadabba.*

24. *Homo sapiens* appeared in Europe approximately _____ years ago.
    a. 20,000
    b. 40,000
    c. 60,000
    d. 80,000
    e. 100,000

25. The "Hobbit People" are thought to have been an offshoot of:
    a. *H. erectus.*
    b. *H. neanderthalensis.*
    c. *H. sapiens.*
    d. *H. habilis.*
    e. *H. ergaster.*

WHAT'S IT ALL ABOUT?

Here's a question to help you pull together what you've learned so far using this text.

Question: Given what you have learned about paleoanthropology, where did the first modern humans originate? How much interaction could they have had with other species of Hominini?

What do I do now?
Remember the drill—decide what the question is asking you to do, collect your evidence from this chapter (and the others you've studied), and write!

CHAPTER 21 VIRUSES, BACTERIA, ARCHAEA, AND PROTISTS: THE DIVERSITY OF LIFE 1

Basic Chapter Concepts

- Although living organisms show extreme diversity of form and function, scientists attempt to group similar organisms into one of three domains: Bacteria, Archaea, and Eukarya.
- Because viruses can replicate only when they take over the replication machinery of another cell, some scientists do not consider them "living" organisms. Thus, viruses define their own category—not "living," not abiotic, and generally bothersome to other organisms.
- Domain Bacteria consist of single-celled, independent-living, haploid organisms that lack all intracellular organelles, including a nucleus. Bacteria, however, exist in every environment and are master decomposer organisms.
- Similar to Bacteria but genetically distinct, Archaea are single-celled organisms that inhabit the most extreme environments on Earth. Some scientists argue that Archaea may also be the oldest organisms on the evolutionary tree.
- The Kingdom Protista is made up of organisms that, although eukaryotic, do not have all of the distinguishing characteristics of plants, animals, or fungi.

CHAPTER SUMMARY

21.1 Life's Categories and the Importance of Microbes
- Microbes are essential to life on Earth. They inhabit an incredible variety of habitats, and they are the most abundant organisms on earth.

21.2 Viruses: Making a Living by Hijacking Cells
- Viruses are little more than nucleic acids, DNA or RNA, encased in a protein coat.
- Viruses inject their nucleic acids into other cells and commandeer the host cell's replication machinery to make more viruses.
- Viruses infect virtually all other living things. Humans respond to viral infection through an immune system that functions to eliminate the viral invaders.

21.3 Bacteria: Masters of Every Environment
- Bacteria are single-celled, asexually reproducing, free-living organisms possessing a single chromosome. Bacteria lack intracellular organelles and a cytoskeleton.
- Bacteria are the primary decomposers in the biosphere. They exist everywhere, including within other organisms. As residents of other organisms, bacteria may also fix nitrogen or produce useful sugars and vitamins.

21.4 Intimate Strangers: Humans and Bacteria
- Most bacteria that live in our bodies have a mutualistic relationship, one in which both organisms benefit. An example is the bacteria that live in our digestive tract to make digestion more efficient.

21.5 Bacteria and Human Disease
- Although only a small number of bacteria are pathogenic, they are responsible for some of humanity's worst diseases.
- Bacteria may also cause disease by producing toxins, or fight disease by producing toxins against other microorganisms.
- The effectiveness of antibiotics is being threatened by overuse, which has led to the emergence of bacterial strains that are increasingly resistant to antibiotics.

21.6 Archaea: From Marginal Player to Center Stage

- Archaea may represent the first type of organism to have evolved on Earth.
- Archaea are similar to Bacteria in that they are free-living, unicellular organisms lacking intracellular organelles. However, Archaea are genetically distinct from both Bacteria and eukaryotic organisms.
- Archaea are frequently found in environments characterized by extremes of temperature, pH, pressure, or salinity.

21.7 Protists: Pioneers in Diversifying Life

- The Kingdom Protista is defined by what its members are not—mostly single-celled organisms that are not plants, not fungi, and not animals.
- Protista acquire energy either through photosynthesis, consumption of other organisms, or decomposition.

21.8 Protists and Sexual Reproduction

- Protista were the first organisms to develop sexual reproduction.

21.9 Photosynthesizing Protists

- The primary example of photosynthesizing protists are the algae; these organisms produce oxygen and are a part of the food webs of many organisms, both terrestrial and aquatic.

21.10 Heterotrophic Protists

- Organisms that are heterotrophic protists consume other organisms or live on decomposing material.
- The protists can be classified by their type of movement: whether they use pseudopods, cilia, or flagella.

WORD ROOTS

auto- = self; **-troph** = nutritive (e.g., an *autotroph* is an organism that synthesizes its own food from simple inorganic compounds)

binary = made up or based on two things or parts; **fission** = a splitting or breaking into parts (e.g., *binary fission* is asexual reproduction by division of one cell or body into two equal or nearly equal parts)

-phile = loving (e.g., an extremo*phile* is an organism, especially an archaean, that lives in conditions of extreme temperature, acidity, alkalinity, or chemical concentration)

hetero- = other, different (e.g., a *hetero*troph is a organism that cannot make its own food and must feed on other plants and animals)

phyto- = plant (e.g., a *phyto*plankton is an aquatic community of floating or weakly swimming photoautotrophs)

KEY TERMS

algae _____

antibiotics _____

autotroph _____

binary fission _____

capsid _____

colonial multicellularity _____

eukaryote _____

extremophile _____

heterotroph _____

mutualism _____

pathogenic _____

phytoplankton _____

prokaryote _____

protist _____

true multicellularity _____

virus _____

FLASH CARDS

To use the flash cards, tear the page from the book and cut along the dotted lines. The key term appears on one side of the flash card, and its definition appears on the opposite side.

algae

capsid

antibiotics

colonial multicellularity

autotroph

eukaryote

binary fission

extremophile

a protein coat that surrounds the genetic material in almost all viruses

protists that perform photosynthesis

a form of life in which individual cells form stable associations with one another but do not take on specialized roles

chemical compounds produced by one microorganism that are toxic to another microorganism

an organism that is a member of Domain Eukarya

any organism that manufactures its own food; almost all plants and algae, and certain bacteria, are autotrophs

an organism that grows optimally in one or more conditions that would kill most other organisms

the form of reproduction carried out by prokaryotic cells, in which the chromosome replicates and the cell pinches between the attachment points of the two resulting chromosomes to form two new cells; in this type of simple cell splitting, each pair of daughter cells is an exact replica of the parental cell

heterotroph

prokaryote

mutualism

protist

pathogenic

true multicellularity

phytoplankton

virus

a single-celled organism whose complement of DNA is not contained within a nucleus

a type of organism that cannot manufacture its own food but must instead get food from elsewhere; animals are heterotrophs

a eukaryotic organism that does not have all the defining characteristics of a plant, animal, or fungus; the catchall term *protist* is used to refer to several distinct evolutionary lines of eukaryotic organisms, most of them single-celled and aquatic

a form of relationship between two organisms in which both organisms benefit

a form of life in which individual cells exist in stable groups, with different cells in a group specializing in different functions

disease-causing; viruses, bacteria, protists, and fungi are spoken of as being pathogenic or nonpathogenic

a noncellular replicating entity that must invade a living cell to replicate itself

small photosynthesizing organisms that drift in the upper layers of oceans or bodies of freshwater, often forming the base of aquatic food webs

SELF TEST

Once you have finished studying this chapter, close your books, grab a pencil, and spend the next 15 to 20 minutes completing this practice test.

Compare and Contrast

For each of the following paired terms, write a sentence of comparison ("Both") and a sentence of contrast ("However,").

autotroph/heterotroph
phytoplankton/zooplankton
virus/protist
colonial multicellularity/true multicellularity
cilia/flagella

Short Answer

1. Describe the principle behind vaccination. What vaccinations have you had?

2. What beneficial roles do bacteria play in the environment?

3. Evolutionarily, where do archaea seem to fit into the scheme of things?

4. What do all protists have in common? List five common attributes.

5. What kind of environment does an extremophile live in? If there were life on Venus, what do you think it would look like? (Keep in mind that Venus is a lot closer to the sun than to Earth.)

6. What are phytoplankton?

7. Which domain is most similar to the Domain Archaea, and why?

8. Why are antibiotics more effective against bacteria than viruses?

9. When did protists evolve?

10. Using a pond as an example, explain the importance of the protists to this type of ecosystem.

Multiple Choice

Circle the letter that best answers the question.

1. Which of the following are the domains of life?
   a. Eukarya, Fungi, Algae
   b. Eukarya, Viruses, Prokaryotes
   c. Eukaryotes, Prokaryotes, Plants
   d. Eukarya, Archaea, Bacteria
   e. Plants, Animals, Fungi

2. Photosynthetic organisms include:
   a. protists.
   b. fungi.
   c. Bacteria.

   d. yeasts.
   e. a and c

3. Which of the following is a protist?
   a. mushroom
   b. whale
   c. worm
   d. alga
   e. fish

4. Sewage treatment plants use:
   a. Bacteria and Archaea.
   b. yeast and Archaea.
   c. bacteria and fungi.
   d. bacteria and protists.
   e. protists and fungi.

5. Bacteria are:
   a. always harmful to humans.
   b. always beneficial to humans.
   c. very large viruses.
   d. generally haploid, unicellular organisms.
   e. never found in nature.

6. Viruses are part of the:
   a. Domain Bacteria.
   b. Domain Archaea.
   c. Kingdom Protista.
   d. Kingdom Viridia.
   e. none of the above

7. Are viruses alive?
   a. No, their lack of replicative ability prevents classification as alive.
   b. No, they don't contain genetic material.
   c. Yes, all pathogens are living organisms.
   d. Yes, they use a protein coat instead of a cell plasma membrane.
   e. Yes, they are capable of very sophisticated behavior and, therefore, must be alive.

8. Bacteria within the human body:
   a. don't exist.
   b. always cause disease.
   c. may help in the digestion of food.
   d. should be treated as pathogenic.
   e. belong to the Domain Archaea.

9. Microscopic organisms (microbes):
   a. produce more than half of the world's oxygen.
   b. provide sources of nitrogen for plants to use.
   c. act as decomposers of organic material.
   d. are vital to the digestion of some animals.
   e. all of the above

10. Instead of sterilizing your water bottles after your weekend camping trip as you usually do, you decide to put the few remaining drops of water on a microscope slide. You observe a diverse group of single-celled creatures. After consulting a few

books, you are pretty certain the organisms are *not* archaea. Why not?

a. Archaea cannot exist in an oxygen atmosphere.

b. Archaea are unlikely to be found in areas where you would go for a weekend camping trip.

c. Archaea are too large to fit on a microscope slide.

d. Archaea are never found in water.

e. Archaea are too small to fit on a microscope slide.

11. You try to determine more about your mysterious creatures. Which of the following attributes will tell you definitively whether they are prokaryotes or eukaryotes?

a. conduct photosynthesis

b. single-celled

c. sexual reproduction

d. presence of DNA

e. whether or not they require oxygen to live

12. Suppose you determine that your creatures are eukaryotic and acquire energy by decomposing organic matter. What type of protist would they be?

a. fungi-like

b. animal-like

c. plant-like

d. bacteria-like

e. viruses

Match the following terms with their description. Each choice may be used once, more than once, or not at all.

a. algae

b. virus

c. protozoan

d. extremophiles

e. Bacteria

13. ____B____ Noncellular infectious agent

14. ____C____ Animal-like protist

15. ____A____ Aquatic protists

16. ____D____ Thrives in high heat environment

17. ____E____ Prokaryotic single-celled organisms

18. How do algae and heterotrophic protists differ?

a. Heterotrophic protists can move, and algae cannot.

b. Algae are free-living, and heterotrophic protists are parasitic.

c. Heterotrophic protists can photosynthesize, and algae cannot.

d. Algae are photosynthetic, whereas heterotrophic protists get their food from consuming other organisms.

e. Algae are prokaryotes, and heterotrophic protists are eukaryotes.

19. A virus is characterized by all but which one of the following?

a. DNA or RNA

b. a protein coat

c. an envelope that surrounds the capsid

d. noncellular organization

e. enzymes of respiration

20. When a virus infects a cell, it forces the cell to manufacture:

a. more mitochondria to provide energy for the virus.

b. more food particles.

c. more genes to incorporate the viral genes.

d. more viral particles.

e. bacteria to kill the virus.

21. Alexander Fleming discovered penicillin when he:

a. found that bacteria were growing in his culture dishes.

b. found that viruses had infected his bacterial cells.

c. found that a fungus was growing in his bacterial culture.

d. accidentally dropped bread mold in his lab.

e. injected someone with fungal spores and their bacterial infection was cured.

22. The simplest of the eukaryotes are:

a. bacteria.

b. plants.

c. animals.

d. protists.

e. fungi.

23. Overuse of antibiotics in medicine and agriculture can be dangerous because:

a. many people are allergic to antibiotics.

b. it may increase the occurrence of antibiotic-resistant bacteria.

c. it may damage natural bacteria found in the soil and air.

d. it may limit the ability of bacteria to produce natural antibiotics.

e. antibiotics are expensive to produce.

24. Heterotrophic protists can move by:

a. cilia.

b. flagella.

c. pseudopods.

d. cellular extensions.

e. all of the above

25. The emergence of sexual reproduction among the protists was important because:

a. it provided increased genetic variation.

b. it led to the emergence of male and female organisms.

c. protists no longer had to depend on binary fission to reproduce.

d. it led to increased numbers of chromosomes.

e. it led to the development of multicellular organisms.

WHAT'S IT ALL ABOUT?

Here's a question to help you pull together what you've learned so far using this text.

Question: Members of the Kingdom Protista seem to defy the orderly patterns of evolution described in Chapter 19, in which a simple ancestral organism evolves into organisms of increasing complexity by adding more features. The Protista, however, seem to be a collection of organisms of diverse size, shape, and lifestyle, some of which are related to the ancestors of modern plants, others related to the ancestors of modern animals, and still others related to the ancestors of modern protists. Why do the organisms in this kingdom persist?

What do I do now?

Remember the drill—decide what the question is asking you to do, collect your evidence from this chapter (and the others you've studied), and write!

CHAPTER 22 FUNGI AND PLANTS: THE DIVERSITY OF LIFE 2

Basic Chapter Concepts

- Despite the great diversity in size, shape, and function, all members of Eukarya possess nucleated, diploid cells that also contain other organelles and structural proteins. Members of the Domain Eukarya are the only organisms that show true multicellularity. Eukarya is divided into four kingdoms: Protista, Fungi, Plantae, and Animalia.

CHAPTER SUMMARY

22.1 The Fungi: Life as a Web of Slender Threads
- Fungi are mostly multicellular and sessile.
- The fungi are not plants, because they are heterotrophic.
- Fungi have hyphae that form a web known as the mycelium.

22.2 Roles of Fungi in Society and Nature
- Fungi are important to the biosphere as decomposers and mineral and water scavengers for other organisms.
- Some fungi are pathogenic.
- Fungi (yeast) are important in food processing.

22.3 Structure and Reproduction in Fungi
- The arrangement of hyphae and septa allows material to flow from one cell to the next. Thus, resources quickly become available for new growth.
- Fungi cells can be dikaryotic, with two separate haploid nuclei. During fusion these become one, making the cell truly diploid.
- Mushroom caps allow the release of spores into the environment. Spores can grow and differentiate into new organisms without fusion.
- Most fungi can reproduce sexually or asexually.

22.4 Categories of Fungi
- There are four major categories of fungi: basidiomycetes, ascomycetes, zygomycetes, and chytrids.
- The basidiomycetes, ascomycetes, and zygomycetes are defined by their reproductive methods. The chytrids are thought to be most similar to ancestral types.

22.5 Fungal Associations: Lichens and Mycorrhizae
- Lichens are the association of a fungus with an alga or bacterium. Fungi provide water, carbon dioxide, minerals, and protection while their symbiotic partners provide an energy source.
- Mycorrhizae are fungi that form symbiotic relationships with plants. Here the plants provide the energy source (via photosynthesis). The vast majority of seed plants form these relationships.

22.6 Plants: The Foundation for Much of Life
- Members of the Kingdom Plantae are mostly sessile, multicellular organisms that can make their own food through the process of photosynthesis.
- Plants are also characterized by the presence of rigid cell walls, which regulate water uptake, and by a life cycle featuring alternating haploid and diploid generations.

22.7 Types of Plants
- There are four main categories of plants: bryophytes, seedless vascular plants, gymnosperms, and angiosperms. These categories are defined by the presence or absence of a vascular system and by method of reproduction, either by spores or by seeds.

22.8 Angiosperm–Animal Interactions
- Flowering plants use animals for sperm (pollen) and seed dispersal.
- Angiosperm seeds contain the nutrient-rich endosperm, which nourishes the developing embryo, and are usually packaged in a food source (fruit) for the dispersing animal.

WORD ROOTS

-phyte = plant (e.g., *phyto*estrogens are estrogen-mimicking compounds that are produced by plants)

-sperm = seed (e.g., gymno*sperm*s are plants with naked seeds)

karyo- = nut, seed, nucleus (e.g., *karyo*kinesis is the stage of the cell cycle that involves the division of the nuclear material)

KEY TERMS

alternation of generations _____

angiosperm _____

bryophyte _____

cell wall _____

chloroplast _____

dikaryotic phase _____

endosperm _____

gametophyte generation _____

gymnosperm _____

hyphae _____

lichen _____

mycelium _____

mycorrhizae _____

seed _____

seedless vascular plant _____

sessile _____

spore _____

sporophyte generation _____

FLASH CARDS

To use the flash cards, tear the page from the book and cut along the dotted lines. The key term appears on one side of the flash card, and its definition appears on the opposite side.

alternation of generations	dikaryotic phase
angiosperm	endosperm
bryophyte	gametophyte generation
cell wall	gymnosperm
chloroplast	hyphae

having two haploid nuclei in one cell; a phase of the life cycle of many fungi

---

a life cycle practiced by plants in which successive plant generations alternate between the diploid sporophyte condition and the haploid gametophyte condition

---

the nutrient tissue that surrounds an angiosperm embryo in the seed; the rice and wheat grains that we eat consist mostly of endosperm

---

a flowering seed plant whose seeds are enclosed within the tissue called fruit; angiosperms are the most dominant and diverse of the four principal types of plants; examples include roses, cacti, corn, and deciduous trees

---

the haploid generation in plants that produces gametes (eggs and sperm); in the flowering plants (angiosperms), this generation is microscopic but gives rise to the visible *sporophyte generation*

---

a plant that lacks a vascular (fluid transport) structure; bryophytes are the most primitive of the four principal varieties of plants; mosses are the most familiar example

---

a seed plant whose seeds do not develop within fruit tissue; one of the four principal varieties of plants, gymnosperms reproduce through wind-aided pollination; coniferous trees, such as pine and fir, are the most familiar examples

---

a relatively thick layer of material that forms the periphery of plant, bacterial, and fungal cells

---

the slender filaments that make up the bulk of most fungi

---

the organelle within plant and algae cells that is the site of photosynthesis

lichen

seedless vascular plant

mycelium

sessile

mycorrhizae

spore

seed

sporophyte generation

a plant that has a vascular (fluid transport) structure but that does not reproduce through use of seeds; one of the four principal varieties of plants; ferns are the most familiar example

a composite organism composed of a fungus and either algae or photosynthesizing bacteria

fixed in location; organisms such as mushrooms and mature sponges are sessile

a web of fungal hyphae that makes up the major part of a fungus

a reproductive cell that can develop into a new organism without fusing with another reproductive cell; the term *spore* also refers to a dormant, stress-resistant form of a bacterial or fungal cell

associations of plant roots and fungal hyphae; the fungal hyphae absorb minerals, growth hormones, and water that are then available to the plant, and the fungus gets carbohydrates from the photosynthesizing plant

the diploid, spore-producing plant generation; this generation is the dominant, visible generation in flowering plants; contrast with gametophyte generation

a reproductive structure in plants that includes a plant embryo, its food supply, and a tough protective casing

SELF TEST

After you have finished studying this chapter, close your books, grab a pencil, and spend the next 15 to 20 minutes completing this practice test.

Compare and Contrast

For each of the following paired terms, write a sentence of comparison ("Both") and a sentence of contrast ("However,").

autotroph/heterotroph
club fungi/imperfect fungi
sporophyte/gametophyte
angiosperm/gymnosperm
lichen/mycorrhizae

Short Answer

1. What is unusual about the sperm of bryophytes?

2. What is a lichen?

3. Describe two adaptations of gymnosperms to living on land.

4. What is the function of nectar?

5. Explain the role that mushroom caps play in reproduction.

6. Why is the production of fruit an indication of a mutually beneficial relationship between plants and animals?

7. Explain how fungi can be both asexual and sexual reproducers.

8. Describe how you have used endosperm in your daily life.

Multiple Choice

Circle the letter that best answers the question.

1. Fungi have formed mutualistic association with all of the following *except:*
   a. protists.
   b. bacteria.
   c. plants.
   d. animals.
   e. all of the above

2. Which of the following is *not* a classification of fungi?
   a. cup fungi
   b. club fungi
   c. bread molds
   d. mushroom
   e. phytofungi

3. Which of the following features is *not* shared by all fungi?
   a. hyphae
   b. heterotrophic
   c. mycelium
   d. eukaryotic
   e. all of these are common characteristics of fungi

4. The dinner conversation with your new sweetheart's family is not going well. To move the discussion along, you start describing the various classifications of fungi. Your sweetie's obnoxious younger sister challenges you to place the mushrooms in the evening's entrée in the correct category. Without hesitation you (correctly) say:
   a. "Imperfect fungi!"
   b. "Cup fungi!"
   c. "Club fungi!"
   d. "Bread mold!"
   e. "Shelf fungi!"

5. Plants use several methods to attract animals to their reproductive structures. Why do they want these pollinators to carry away pollen after their visit?
   a. Pollen contains the plant sperm for reproduction.
   b. Pollen contains the toxic by-products of photosynthesis.
   c. Fungal plant invaders force the plant to produce pollen to distribute their spores.
   d. Bacteria tend to infest the plant parts known as pollen, and the plant is trying to remove them.
   e. Bacteria tend to infest the plant parts known as pollen and the plant is trying to remove them.

6. All of these are shared common features of plants *except:*
   a. photoautotrophic.
   b. multicellular.
   c. mobile.
   d. eukaryotic.
   e. cell wall.

7. A common form of seasonal allergy results from the release of pollen by plants. Which of the following would *not* be responsible for your itchy, watery eyes?
   a. bryophytes
   b. vascular seedless plants
   c. gymnosperms
   d. angiosperms
   e. a and b

8. Which of the following correctly matches the classification with the example?
   a. bryophyte—moss
   b. seedless vascular plant—cactus
   c. gymnosperm—fern
   d. angiosperm—pine tree
   e. gymnosperm—water lily

9. Which of the following most closely resembles the earliest plants?
   a. bryophytes
   b. seedless vascular plants
   c. gymnosperms
   d. angiosperms
   e. fungi

10. You find a new plant species with vascular bundles, but no flowers. How would you classify it?
    a. bryophyte
    b. fern
    c. gymnosperm
    d. angiosperm
    e. b and c

11. The plant life cycle is characterized by alternation of generations. The generation that produces spores is known as the _____ generation, whereas the one that produces the seeds is the _____ generation.
    a. sporophyte; gametophyte
    b. haploid; diploid
    c. bryophyte; angiosperm
    d. neophyte; gametophyte
    e. parental; offspring

12. Your ship crash-lands on planet X2-Alpha, and all of your medical supplies are destroyed. Luckily, X2-Alpha has many of the flora common to Earth. Which of the groups below would be a good likely source for potent antibiotics such as penicillin?
    a. bread molds
    b. club fungi
    c. cup fungi
    d. animal-like protists
    e. archaea

13. Being dikaryotic means:
    a. cells can be diploid while having haploid nuclei.
    b. having both a sporophyte and gametophyte generation.
    c. having both flowers and seeds.
    d. being able to reproduce sexually and asexually.
    e. having cells that are resistant to desiccation.

14. Which of the following best describes chloroplasts?
    a. containing two haploid nuclei
    b. form associations with plant roots
    c. site of photosynthesis
    d. multicellular, vascular system
    e. multicellular, vascular system, seeds

15. Which of the following best describes the gymnosperms?
    a. containing two haploid nuclei
    b. form associations with plant roots
    c. site of photosynthesis
    d. multicellular, vascular system
    e. multicellular, vascular system, seeds

16. Which of the following best describes mycorrhizae?
    a. containing two haploid nuclei
    b. form associations with plant roots
    c. site of photosynthesis
    d. multicellular, vascular system
    e. multicellular, vascular system, seeds

17. Which of the following best describes a dikaryotic organism?
    a. containing two haploid nuclei
    b. form associations with plant roots
    c. site of photosynthesis
    d. multicellular, vascular system
    e. multicellular, vascular system, seeds

18. Which of the following best describes ferns?
    a. containing two haploid nuclei
    b. form associations with plant roots
    c. site of photosynthesis
    d. multicellular, vascular system
    e. multicellular, vascular system, seeds

19. Which of the following is *not* a kingdom of the Eukarya?
    a. Fungi
    b. Moldi
    c. Plantae
    d. Animalia
    e. Protista

20. What is the major difference between fertilization in humans and fusion in fungi?
    a. In humans, there are two sources of DNA; fungi are self-fertilizing.
    b. In humans, eggs can self-fertilize with polar bodies; in fungi, two nuclei are required.
    c. In humans, fertilization is external; in fungi, fertilization is internal.
    d. In humans, two nuclei join to form a dikaryote; in fungi, the nuclei merge to form one.
    e. In humans, two nuclei join to form one; in fungi, the nuclei remain intact, forming a dikaryote.

21. Which of the following are the correct two components of a lichen?
    i. fungus
    ii. bacteria
    iii. algae
    iv. plant
       a. i, ii
       b. i, iii

c. i, ii, iv

d. ii, iii, iv

e. i, ii, iii

22. What is the function of endosperm?

    a. to produce a smell that repels predators

    b. to produce a smell that attracts seed dispersers

    c. to provide energy for flower production

    d. to provide sperm to fertilize eggs

    e. to nurture the plant embryo

23. Why do some plants have flowers?

    a. Plants without vascular systems rely on flowers to collect water.

    b. Plants in poor soils rely on flowers to catch mineral-rich insect droppings.

    c. Plants in sunny environments use flowers to maximize sun exposure for photosynthesis.

    d. Plants adapted to land rely on flowers to attract pollinators.

e. Plants produce flowers when infected with certain fungi.

24. Which of the following is a major division of fungi?

    a. basidiomycetes

    b. zygommaticus

    c. chytrids

    d. acetomycetes

    e. hyphenae

25. Which of following has not been attributed to fungi?

    a. athlete's foot

    b. ringworm

    c. jock itch

    d. corn smut

    e. herpes

WHAT'S IT ALL ABOUT?

Here's a question to help you pull together what you've learned so far using this text.

Question: As evolutionary adaptations go, the development of angiosperms seems like a risky proposition. These plants depend on the existence of some other organism to complete their life cycle. Early angiosperms did not "know" that animals existed that would disperse their pollen or seeds, so how could such a dependence develop?

What do I do now?

Remember the drill—decide what the question is asking you to do, collect your evidence from this chapter (and the others you've studied), and write!

CHAPTER 23   ANIMALS: THE DIVERSITY OF LIFE 3

Basic Chapter Concepts

- Animals are a remarkably diverse collection of multicellular, heterotrophic organisms. The single feature uniquely defining all animals is that they pass through a blastula stage during development.
- The most primitive animals lack a central body cavity, a coelom. Sponges (Porifera) and jellyfish (Cnidaria) are examples of noncoelomate animals.
- More complex animals that possess a coelom can be further differentiated by the orientation of their nerve cord. Protostomes have a ventral nerve cord, while deuterostomes have a dorsal nerve cord. Protostomes and deuterostomes define two branches of equally complex animals on the family tree.

CHAPTER SUMMARY

23.1 What Is an Animal?
- Animals are classified according to body plan into phyla (singular, *phylum*). Body plans become increasingly complex as we move up the family tree because of the addition of body features.
- The most primitive group of animals lacks a body plan—the sponges (Porifera) do not have tissues, symmetry, or a body cavity.
- More complex animal bodies have tissues organized into organs and show some type of body symmetry, either radial or bilateral.
- The most complex animals possess an internal body cavity, a coelom, which provides protection to internal organs and flexibility to the body.
- Coelomate animals are further categorized based on whether the notochord orients to the ventral or dorsal side of the body.

23.2 Animal Types: The Family Tree
- Animals are classified into phyla; each phylum consists of animals that share basic body structures.
- All animals pass through a blastula stage of development; the blastula later invaginates to form the digestive cavity. In protostomes, the opening to the invagination becomes the mouth; in deuterostomes, the opening becomes the anus.
- All protostomes possess tissues organized into organs, including nerve cells, bilateral symmetry, and a layer of tissue in the embryo, called the mesoderm, which allows for the development of more complex tissues. Examples of protostomes include flatworms, roundworms, annelids, molluscs, and arthropods. Examples of deuterostomes include echinoderms and chordates.

23.3 Phylum Porifera: The Sponges
- Sponges are the odd members of the animal kingdom, not even possessing tissues; but a sponge is still a cooperative entity. The independent cells work together to feed by filtering the water that passes through the sponge.

23.4 Phylum Cnidaria: Jellyfish and Others
- Cnidarians—jellyfish, corals, anemones, and hydrozoans—possess a sac-like body with a single opening to a gastrovascular cavity that carries out both digestion and nutrient transport.
- Cnidarians have rudimentary nervous and muscle systems that allow the animals to move by expelling a jet of water.

23.5 Phylum Platyhelminthes: Flatworms
- Flatworms (Platyhelminthes) are the most primitive protostomes, lacking a coelom but possessing a collection of nerve cells and primitive eyes at one end, defining a "head." Most flatworms are parasitic species.

23.6   Phylum Annelida: Segmented Worms

- Body segmentation appears in the Phylum Annelida (segmented worms). Segmentation allows parts of the body to function independently, giving flexibility and strength to the body plan.
- Earthworms, leeches, and clam worms are examples of annelids.

23.7   Phylum Mollusca: Snails, Oysters, Squid, and More

- With the Phylum Mollusca, we first see a true circulatory system, an open system in which blood is pumped into the tissues and diffuses back to the veins, as well as a stomach and kidney.
- One group of molluscs (the cephalopods) shows signs of intelligence, indications of a developed nervous system. The one feature shared by all molluscs is the mantle, a protective tissue layer on the upper surface of the body.

23.8   Phylum Nematoda: Roundworms

- Phylum Nematoda (roundworms) are the first protostomes to have a coelom, so their bodies have two openings (mouth and anus). Most roundworms are also parasitic species.

23.9   Phylum Arthropoda: Insects, Lobsters, Spiders, and More

- The most numerous protostomes, and most numerous types of animals, are the members of Phylum Arthropoda, which includes 1,000,000 species of insects.
- All arthropods are characterized by an exoskeleton and paired, jointed appendages. In addition to the insects (Subphylum Uniramia), other members of this very large phylum include spiders, ticks, and horseshoe crabs (Subphylum Chelicerata); and crabs, lobsters, and shrimp (Subphylum Crustacea).

23.10   Phylum Echinodermata: Sea Stars, Sea Urchins, and More

- The Echinodermata appear to be a "throwback" to the more primitive protostomes; sea stars lack a brain, have primitive eyespots and radial symmetry, and can regenerate from part of the central disc and an arm. But echinoderm embryos have bilateral symmetry, a feature that makes these organisms more developed than the Cnidaria.

23.11   Phylum Chordata: Mostly Animals with Backbones

- Vertebrate animals, including humans, are members of the Phylum Chordata.
- All chordates possess a notochord, a dorsal nerve cord, pharyngeal slits, and a post-anal tail, although some members of the phylum, like us, show some of these features only during embryonic development.
- Fifty percent of the chordate species are fish; and among fish, the most common species are ray-finned, bony fishes. However, it was the lobe-finned fishes that served as the ancestor of all land vertebrates.
- Amphibians, such as frogs, are terrestrial vertebrates that must spend some part of their life cycle in an aquatic environment.
- Reptiles and birds, as different as they appear, probably evolved from a common, dinosaur ancestor. These two kinds of animals encase their embryos in an amniotic egg, a structure that protects the developing embryo from drying out.
- Only a small number of chordates are mammals, but these animals (like us) have had a profound effect on the Earth. All mammals have mammary glands, providing milk for the young. They can maintain a constant body temperature, which means they can inhabit any environment, including cold ones.
- Three classes of mammals are distinguished by modes of reproduction. Monotremes are egg-layers. In the marsupials, development of the young animal occurs both inside and outside the mother's body. The young of placental mammals develop only within their mother's bodies, depending on the mother's blood supply for nourishment.

WORD ROOTS

bi- = two; **lateral** = of or relating to the side (e.g., *bilateral* symmetry is a body plan in which the main axis divides the body into two halves that are mirror images of one another)

endo- = within, inside (e.g., *endo*thermic refers to metabolically generated heat that keeps an animal warm)

exo- = outside (e.g., an *exo*skeleton is an external skeleton, as found in arthropods)

meta- = change, transformation (e.g., *meta*morphosis refers to major changes in the body form of certain animals, as, for example, from tadpole to frog)

tetra- = four; **-pod** = foot (e.g., a *tetrapod* is a four-legged vertebrate; e.g., horse)

KEY TERMS

amniotic egg _____

bilateral symmetry _____

bivalves _____

body segmentation _____

cephalopods _____

closed circulation system _____

coelom _____

dorsal nerve cord _____

ectothermic _____

endothermic _____

exoskeleton _____

gastropods _____

hermaphroditic _____

invertebrate _____

mammary glands _____

marsupial _____

metamorphosis _____

molting _____

monotreme _____

notochord _____

open circulation system _____

organ _____

oviparous _____

paired, jointed appendages _____

parasite _____

pharyngeal slits _____

phylum _____

placenta _____

placental mammal _____

post-anal tail _____

radial symmetry _____

swim bladder _____

symmetry _____

tetrapods _____

tissue _____

vertebral column _____

viviparous _____

FLASH CARDS

To use the flash cards, tear the page from the book and cut along the dotted lines. The key term appears on one side of the flash card, and its definition appears on the opposite side.

amniotic egg	dorsal nerve cord
bilateral symmetry	ectothermic
bivalves	endothermic
body segmentation	exoskeleton
cephalopods	gastropods
closed circulation system	hermaphroditic
coelom	invertebrate

a rod-shaped dorsal structure consisting of nerve cells, running from the chordate animal's head to its tail

an egg with a hard outer casing and an inner series of membranes and fluids that provide protection, nutrients, and waste disposal for a growing embryo; the evolution of the amniotic egg, in reptiles, freed them from the constraint of having to reproduce near water

having an internal temperature that is controlled largely by the temperature of the external environment; for example, lizards are ectothermic and often bask in the sun to warm up

a bodily symmetry in which opposite sides of a sagittal plane are mirror images of one another; animals generally are bilaterally symmetrical

maintaining a relatively stable internal body temperature through use of heat that is generated internally by the organism's own metabolism; mammals are endothermic

a class of molluscs that includes mussels, clams, and oysters, among other organisms

an external material covering the animal body, providing support and protection

a repetition of body parts in an animal; an example can be found in the vertebrae that make up the human vertebral column or backbone

a class of molluscs that includes snails and slugs, among other organisms

a class of molluscs that includes squid, octopus, nautilus, and cuttlefish

a state in which one animal possesses both male and female sex organs

a type of circulatory system, found in all vertebrates and in some invertebrates, in which blood stays within vessels

an animal without a vertebral column

a central body cavity, found in animals, that is lined with cells of mesodermal origin

mammary glands	**open circulation system**
marsupial	**organ**
metamorphosis	**oviparous**
molting	**paired, jointed appendages**
monotreme	**parasite**
notochord	**pharyngeal slits**

a type of circulation, found in many invertebrates, in which arteries carry blood into open spaces called sinuses; the blood then bathes surrounding tissues and is channeled into veins, through which it flows back to the heart

a set of glands that, in female mammals, provide milk for the young

a highly organized unit within an organism, performing one or more functions, that is formed of several kinds of tissue; kidneys, heart, lungs, and liver are all familiar examples of organs in humans

a type of mammal in which the young develop within the mother to a limited extent, inside an egg having a membranous shell; early in development, the egg's membrane disappears, after which the mother delivers a developmentally immature but active marsupial; kangaroos are one example of a marsupial

a condition, seen in all birds and many reptiles, in which fertilized eggs are laid outside the mother's body and then develop there

a change in form in an organism as it develops from an embryo into an adult; common in insects and amphibians

appendages, such as legs, that come in pairs and have joints; such appendages are characteristic of all arthropods

a periodic shedding of an old skeleton followed by the growth of a new one; practiced by animals in the Phyla Nematoda and Arthropoda

organisms that feed off their prey but do not kill them, at least not immediately

egg-laying mammals, represented by the duck-billed platypus and spiny anteaters found in Australia

in animals, openings to the pharyngeal cavity; all chordates possess pharyngeal slits at some point in their development

a dorsal, rod-shaped support organ that exists in embryonic development in all vertebrates and in the adults of some vertebrates

phylum

symmetry

placenta

tetrapods

placental mammal

tissue

post-anal tail

vertebral column

radial symmetry

viviparous

swim bladder

an equivalence of size, shape, and relative position of parts across a dividing line or around a central point

a category of living things, directly subordinate to the category of kingdom, whose members share traits as a result of shared ancestry

any four-limbed vertebrate; all amphibians, reptiles and mammals

a complex network of maternal and embryonic blood vessels and membranes that develops in mammals in pregnancy; the placenta allows nutrients and oxygen to flow to the embryo from the mother, while allowing carbon dioxide and waste to flow from the embryo to the mother

an organized assemblage of similar cells that serves a common function; nervous, epithelial, and muscle tissue are some familiar examples

a type of mammal that is nurtured before birth by the placenta, a network of maternal and embryonic blood vessels and membranes; embryonic placental mammals derive their nutrition not from food stored in an egg, but directly from the mother's circulation

a flexible column of bones extending from the anterior to posterior end of an animal; also known as a backbone, the vertebral column distinguishes vertebrates from other chordates

a tail, existing at some point in the development of all chordates, that is located posterior to the anus

a condition in animals in which fertilized eggs develop inside a mother's body

a type of animal symmetry in which body parts are distributed evenly about a central axis; sea stars are radially symmetrical

an inflatable organ in a fish that the fish can fill with gas to maintain neutral buoyancy—to float in place without expending energy

SELF TEST

Once you have finished studying this chapter, close your books, grab a pencil, and spend the next 15 to 20 minutes completing this practice test.

Compare and Contrast

For each of the following paired terms, write a sentence of comparison ("Both") and a sentence of contrast ("However,").

    roundworms/flatworms
    ectothermy/endothermy
    triploblastic/diploblastic
    bilateral/radial symmetry
    oviparous/viviparous

Short Answer

1. Are deuterostomes more complex animals than protostomes? Why or why not?

2. Lacking any body plan, tissues, organs, or other features of higher organisms, how do the inner cells of a sponge get the food and oxygen they need?

3. Jellyfish are squishy and mobile; corals are hard and sessile. How can these organisms be part of the same phylum (what feature do they share)?

4. Organization of tissues into organs increases the complexity and functionality of the organism. Why do the protostomes have organs but the Cnidaria and Porifera do not?

5. Why are flatworms flat?

6. How do arthropods grow, given that their bodies are covered by an exoskeleton?

7. Survival is closely linked to the ability of an organism to secure enough food. How have insects been able to ensure a food supply?

8. Why would the development of jaws in ancient chordates be a significant advantage to the animal?

9. Why does the amniotic egg represent an evolutionary advantage to the reptiles and birds, compared to the eggs of amphibians?

10. Being able to maintain a constant body temperature (endothermy) is a great advantage to mammals, but it comes at a cost. What do mammals need to be endothermic?

11. What separates the Vertebrata from the other two subphyla in Chordata, the Cephalochordata and the Urochordata?

Multiple Choice

Circle the letter that best answers the question.

1. Your friend Joe spent the summer visiting a number of exotic places and came home with a parasitic infection. In fact, he has been diagnosed as suffering from trichinosis. What phylum is the source of this infection?
   a. Annelida
   b. Platyhelminthes
   c. Mollusca
   d. Nematoda
   e. Porifera

2. Which of the following is *not* a type of reptile?
   a. snakes and lizards
   b. turtles
   c. amphibians
   d. crocodiles and alligators
   e. dinosaurs

3. Most animal fossils involve only the hardest structures of organisms. Which of the following phyla is/are therefore most likely to leave behind such evidence?
   a. Chordata
   b. Arthropoda
   c. Nematoda
   d. Annelida
   e. a and b

4. You take a job as a research assistant at one of the richest fossil fields ever to be discovered. On your first day, you discover a new species, *Neko borgus*, which bears similarities to the modern domesticated cat. *Neko borgus* belongs to which of the following groups?
   a. Chordata
   b. protostomes
   c. acoelomates
   d. Arthropoda
   e. cephalochordates

5. What group comprises more than half of the vertebrate species?
   a. fish
   b. amphibians
   c. mammals
   d. birds
   e. reptiles

6. Which of the following is *not* a defining feature of the Arthropoda?
   a. exoskeleton
   b. jointed appendages

c. notochord
d. protostome
e. coelom

7. What is the fundamental difference between proto-stomes and deuterostomes?
   a. bilateral symmetry
   b. the fate of the blastopore
   c. the presence of tissues
   d. the presence of a notochord
   e. the presence of a coelom

8. Which of the following does *not* belong to the Phylum Mollusca?
   a. gastropods
   b. nematodes
   c. cephalopods
   d. octopus
   e. bivalves

9. Monotremes are:
   a. molluscs.
   b. amphibians.
   c. porifera.
   d. mammals.
   e. annelids.

10. Mammals:
    a. are endothermic.
    b. are coelomates.
    c. have mammary glands.
    d. are vertebrates.
    e. are all of the above

11. Which feature(s) would be shared by both a sea star and a human but *not* a cockroach?
    a. Both are protostomes.
    b. Both are deuterostomes.
    c. Both lack a coelom.
    d. Both are mammals.
    e. Both are chordates.

12. A friend brings you a jar of seawater collected on a recent vacation at the beach. Within the jar you find a blob of what appears to be living tissue, which you eventually determine to be a new kind of animal. You are confident in your conclusion because it fits all of the characteristics of an animal. All animals have the following characteristics *except:*
    a. symmetry.
    b. cells without cell walls.
    c. the blastula stage of development.
    d. heterotrophy.
    e. multicellularity.

Match the following terms with their description. Each choice may be used once, more than once, or not at all.
   a. reptiles
   b. Porifera
   c. Mollusca

d. marsupials
e. Platyhelminthes

13. _____ Lacking symmetry and tissues

14. _____ Tapeworms and other flatworms

15. _____ Squid

16. _____ Turtles, snakes, and lizards

17. _____ Kangaroos

18. Which of the following includes the largest number of species?
    a. animals that are segmented
    b. animals with radial symmetry
    c. animals with a body cavity
    d. animals that are unsegmented
    e. animals with a backbone

19. Which of the following animals is thought to be most closely related to you?
    a. sea star
    b. snail
    c. earthworm
    d. jellyfish
    e. ant

20. Which of the following is *not* a protostome?
    a. earthworm
    b. crayfish or lobster
    c. sea star
    d. squid
    e. earthworm

21. Molting in arthropods involves a change in:
    a. body form and maturity.
    b. sex.
    c. body size.
    d. eating habits.
    e. sex.

22. The notochord is most closely associated with the:
    a. nervous system.
    b. spinal cord.
    c. skeletal system.
    d. skin.
    e. lungs.

23. Birds differ from earlier vertebrates by:
    a. their lack of scales.
    b. the land egg.
    c. the ability to maintain a constant body temperature.
    d. the ability to fertilize eggs internally.
    e. the number of appendages.

24. All but which of the following have cartilaginous skeletons?
    a. sharks
    b. lampreys
    c. perch
    d. rays
    e. skates

25. The most unusual feature of the echinoderms is:
- **a.** radial symmetry.
- **b.** a hard exoskeleton.
- **c.** the lack of a brain.
- **d.** well-developed sense organs.
- **e.** a unique system for locomotion involving tube feet.

WHAT'S IT ALL ABOUT?

Here's a question to help you pull together what you've learned so far using this text.

Question: The family tree of the Kingdom Animalia groups animals according to increasing complexity of form. Think back to Chapter 1 (reread the section on the hierarchy of life to refresh your memory) and the hierarchy of life from molecules to biosphere. How does the hierarchy described by phylogenetics fit into this larger hierarchy of living things?

What do I do now?

Remember the drill—decide what the question is asking you to do, collect your evidence from this chapter (and the others you've studied), and write!

CHAPTER 24 THE ANGIOSPERMS: AN INTRODUCTION TO FLOWERING PLANTS

Basic Chapter Concepts

- Flowering plants vary in their anatomy and life cycles but share common cell types, transport systems, reproductive mechanisms, and growth patterns.
- Plant tissues are made up of three different kinds of cells—parenchyma, sclerenchyma, and collenchyma. These three cell types comprise the four tissue types (dermal, vascular, meristematic, and ground) of all angiosperms.
- All angiosperms reproduce sexually through an alternation of generations.

CHAPTER SUMMARY

24.1 The Importance of Plants
- Flowering plants can be classified by life span and by the number of embryonic leaves. Plants may complete their life cycles annually or biennially, or they may live for many years.
- Important food crops have a single embryonic leaf and are classified as monocotyledons. About 75 percent of all flowering plants are dicotyledons—having two embryonic leaves.

24.2 The Structure of Flowering Plants
- Plant structures can be functionally divided into two groups: roots and shoots. Roots provide support and are responsible for the absorption of water and minerals from the soil. Shoots support the leaves used in photosynthesis and the flowers for reproduction.
- Stems and roots can be modified for storage as well as support.

24.3 Basic Functions in Flowering Plants
- Plant adaptations to living on land include those for reproduction, fluid transport, growth, and defense.
- Plants produce spores that grow into a distinct generation, producing gametophyte plants that look different from the parental plant. However, many of the sporophyte generation stay associated with the parental plants, so it is hard to distinguish these plants from their parents.
- Gametophyte plants, the spore generation, produce haploid gametes (through meiosis) that fuse, producing a diploid zygote that will grow into a sporophyte plant.
- Flowers are the reproductive parts of the gametophyte generation, producing male (pollen grains) and female (eggs) gametes. Fusion of male and female gametes produces an embryo that becomes surrounded by fruit and a seed coat, both derived from the plant ovary.
- Herbaceous plants exhibit only primary growth—vertical growth from the tips of the roots and shoots. Woody plants, those that develop bark, exhibit both primary and secondary growth; secondary growth increases the plant's girth.
- Dermal tissue controls the plant's interactions with the outside world, regulating water and gas exchange, preventing infections, and secreting defense chemicals.
- Ground tissue, the most abundant tissue, is a prime site of photosynthesis.
- Vascular tissue, xylem and phloem, regulates the movement of water and food, respectively.
- Meristematic tissue is the source of the vascular, ground, and dermal tissues because it is the only tissue capable of dividing, growing, and differentiating.
- Plant hormones and cooperative relationships have also allowed plants to effectively utilize their habitats.

24.4 Responding to External Signals
- Plants can respond to external stimuli such as gravity, light, or the presence of nearby structures by differential growth of the stem. These movements are respectively known as gravitropism, phototropism, and thigmotropism.
- Plants can respond to seasonal changes in water availability by changing their production of new structures, such as leaves. Plants that drop their leaves during dry periods are known as deciduous plants.
- Plants coordinate growth and reproduction with the change in light availability across the seasons. This is known as photoperiodism.

WORD ROOTS

photo- = light (e.g., *photo*tropism is the bending of a plant toward a light source)

-tropic, tropism = action or movement (e.g., gravi*tropism* is the movement of a plant in response to a gravity source)

-spore = reproductive structure (e.g., the micro*spore* will develop into a pollen grain)

KEY TERMS

angiosperm _____

anther _____

apical dominance _____

blade _____

carpel _____

deciduous _____

dormancy _____

fibrous root system _____

filament _____

fruit _____

gametophyte generation _____

gravitropism _____

hormone _____

megaspore _____

microspore _____

mycorrhizae _____

nutrient _____

ovary _____

petal _____

petiole _____

phloem _____

photoperiodism _____

phototropism _____

pollen tube _____

root hair _____

seed _____

sepal _____

sperm cell _____

sporophyte generation _____

stamen _____

stigma _____

stomata _____

style _____

taproot system _____

thigmotropism _____

tissue _____

transpiration _____

tube cell _____

vegetative reproduction _____

xylem _____

FLASH CARDS

To use the flash cards, tear the page from the book and cut along the dotted lines. The key term appears on one side of the flash card, and its definition appears on the opposite side.

angiosperm	fibrous root system
anther	filament
apical dominance	fruit
blade	gametophyte generation
carpel	gravitropism
deciduous	hormone
dormancy	megaspore

a plant root system that consists of many roots, all about the same size

---

a flowering seed plant whose seeds are enclosed within the tissue called fruit; angiosperms are the most dominant and diverse of the four principal types of plants; examples include roses, cacti, corn, and deciduous trees

---

the part of a stamen (male reproductive part of a flower) that is shaped like and functions as a stalk and has an anther at the top

---

the part of a flower that produces pollen grains; the anther is on top of a filament, and together they make up the flower's stamen

---

the mature ovary of any flowering plant; many fruits protect the underlying seeds, and many attract animals that will eat the fruit and disperse the seeds

---

suppression of the growth of the lateral branches of a plant through the activity of apical meristems

---

the haploid generation in plants that produces gametes (eggs and sperm); in the flowering plants (angiosperms), this generation is microscopic, but gives rise to the visible *sporophyte generation*

---

in plants, the major, broad part of a leaf

---

the bending of a plant's roots or shoots in response to gravity; this capability helps a plant orient its roots and shoots properly—roots toward the center of the earth, shoots away from it

---

the female reproductive structure of a flower, consisting of an ovary, a style, and a stigma

---

a substance that, when released in one part of an organism, goes on to prompt physiological activity in another part of the organism; both plants and animals have hormones

---

refers to plants that show a coordinated, seasonal loss of leaves; this strategy allows plants to conserve water during a time they could perform little photosynthesis anyway

---

in plants, the first cell in the female gametophyte generation; this cell eventually develops into the seven-celled embryo sac, which is the mature female gametophyte plant

---

a state in which growth is suspended and there is a prolonged low level of metabolic activity; dormancy allows organisms to conserve energy during times of unfavorable environmental conditions

microspore	photoperiodism
mycorrhizae	phototropism
nutrient	pollen tube
ovary	root hair
petal	seed
petiole	sepal
phloem	sperm cell

the ability of a plant to respond to changes it experiences in the daily duration of darkness relative to light

in plants, the first cell in the male gametophyte generation; the microspore will develop into a pollen grain that consists at maturity of a tube cell, two sperm cells, and a protective coat

the bending of a plant's shoots in response to light; generally, this capability helps a plant grow toward the sun to get the most available sunlight

associations of plant roots and fungal hyphae; the fungal hyphae absorb minerals, growth hormones, and water, which are then available to the plant, and the fungus gets carbohydrates from the photosynthesizing plant

a tube-like structure that sprouts from a pollen grain that has landed on the stigma of a plant; the pollen tube grows down toward the egg, and sperm cells then move down through the pollen tube

a substance found in food that does at least one of three things: provides energy, provides a structural building block, or regulates a physical process; there are six classes of nutrients: water, minerals and vitamins; and carbohydrates, lipids, and proteins

in plants, a thread-like extension of a root cell; root hairs greatly increase the surface area of roots, thus allowing greater absorption of water and nutrients

in flowering plants, the area, located at the base of the carpel, where fertilization of the egg and early development of the embryo occur; in animals, the female reproductive organ in which eggs develop

a reproductive structure in plants that includes a plant embryo, its food supply, and a tough protective casing

the colorful, leaf-like structure of a flower; petals attract pollinators

the leaf-like structure that, with other sepals, protects the flower before it opens

the stalk of a leaf, attaching it to a branch or trunk

in flowering plants, either of two cells in a pollen grain, one of which fertilizes an egg, the other of which fertilizes the central cell in an embryo sac; in animals, the male gamete, which fertilizes the female gamete (the egg)

in vascular plants, the fluid-transporting tissue that conducts the food produced in photosynthesis along with some hormones and other compounds

sporophyte generation

thigmotropism

stamen

tissue

stigma

transpiration

stomata

tube cell

style

vegetative reproduction

taproot system

xylem

the growth of a plant in response to touch; this capability allows tendrils to wrap around other objects, thus helping a plant climb upward toward light

the diploid, spore-producing plant generation; this generation is the dominant, visible generation in flowering plants; contrast with *gametophyte generation*

an organized assemblage of similar cells that serves a common function; nervous, epithelial, and muscle tissue are some familiar examples

the male reproductive structure in a flower, consisting of a stalk-like filament topped by a pollen-producing anther

the process by which plants lose water when water vapor leaves the plant through open stomata; more than 90 percent of the water that enters a plant evaporates into the atmosphere via transpiration

in flowering plants, the tip end of a flower's carpel, where pollen grains are deposited prior to fertilization

a cell in a pollen grain that, following arrival of the pollen grain on a stigma, germinates and forms a pollen tube that conducts sperm cells to the embryo sac

microscopic pores, found in greatest abundance on the undersides of leaves, that allow plants to exchange gases with the atmosphere; carbon dioxide moves into plants through the stomata, while oxygen and water vapor move out

form of asexual reproduction practiced by plants in which a portion of an initial plant can grow into a second plant; severed "runners" of aspen trees can grow into separate trees through vegetative reproduction

in flowering plants, the stalk-like extension of a carpel that connects its stigma and ovary

the tissue through which water and dissolved minerals flow in vascular plants

the type of plant root system that consists of a large central root and many smaller lateral roots

SELF TEST

After you have finished studying this chapter, close your books, grab a pencil, and spend the next 15 to 20 minutes completing this practice test.

Compare and Contrast

For each of the following paired terms, write a sentence of comparison ("Both") and a sentence of contrast ("However,").

pollen/endosperm
phototropism/thigmotropism
primary/secondary growth
cork cambium/vascular cambium
petal/petiole

Short Answer

1. Explain the significance of this statement: "An ovary a day keeps the doctor away."

2. What is the primary function of meristem?

3. Explain how transpiration directs the flow of xylem in plants.

4. What is the adaptive benefit of thigmotropism?

Multiple Choice

Circle the letter that best answers the question.

1. Secondary xylem and secondary phloem form which of the following?
   a. epididymis
   b. bark
   c. gametophytes
   d. chloroplasts
   e. flowers

2. What type of cell forms the bulk of a herbaceous plant and carries out a variety of functions in that plant?
   a. collenchyma
   b. parenchyma
   c. meristems
   d. vascular bundles
   e. sclerenchyma

3. Which meristematic tissues are responsible for primary plant growth?
   a. vascular cambium
   b. cork cambium
   c. apical meristem
   d. vascular meristem
   e. stomatal meristem

4. As much as _____ of the calories that human beings consume comes from plants.
   a. 10%
   b. 20%
   c. 40%
   d. 75%
   e. 90%

5. Which of the following is *not* a classification of plants?
   a. bryophyte
   b. angiosperm
   c. gymnosperm
   d. basidiophyte
   e. seedless vascular plants

6. Within the plant kingdom, angiosperms make up at least _____ of species.
   a. 20% to 30%
   b. 40% to 50%
   c. 60% to 70%
   d. 80% to 90%
   e. 100%

7. Plant root tissues are used for all of the following *except:*
   a. water absorption.
   b. mineral absorption.
   c. photosynthesis.
   d. support.
   e. food storage.

8. A mature pollen grain consists of all of the following *except:*
   a. an outer coat of ground tissue.
   b. two sperm cells.
   c. two egg cells.
   d. one tube cell.
   e. All of the above are part of the pollen grain.

9. How do sperm reach the ovule?
   a. by entering through leaf stomata
   b. by entering through the root cortex
   c. by the pollen tube
   d. through ingestion by an animal pollinator
   e. by chloroplastic immigration

10. If the surrounding air was saturated with water (100% humidity), what would happen to xylem flow?
    a. It would move faster.
    b. It would reverse direction.
    c. It would slow or stop.

d. Water content of air has nothing to do with xylem flow.

e. The xylem would take on water.

11. After a 10-year absence, you return to the house where you grew up. Outside in the yard, you find that your old tree house is still in good shape. You decide to climb the ladder nailed into the side of the oak tree one last time to reminisce. You notice that 10 years of secondary growth has caused the steps of the ladder to:

a. move downward (toward the ground).

b. move upward (away from the ground).

c. move closer to the center of the trunk.

d. move farther from the center of the trunk.

e. not move at all.

12. Which of the following is *not* an angiosperm?

a. cactus

b. cotton plant

c. fern

d. rose bush

e. oak tree

13. If you accidentally chop off the top of your favorite plant, will it continue to grow?

a. No, the apical meristem has been removed.

b. No, the vascular system has been disrupted, so the plant will die.

c. Yes, new apical meristem will form immediately.

d. Yes, new vascular tissue will form as needed.

e. Yes, but not from the top; other meristematic tissue will take over.

14. Why does a plant die if you remove a ring of its bark?

a. Removing the bark will expose the underlying tissue to air, which dries it out.

b. Removing the bark will expose the plant to bacterial infection.

c. Removing the bark will disrupt the photosynthesis machinery.

d. Removing the bark will disrupt the vascular tissue.

e. Removing the bark will prevent the plant from forming flowers.

15. What causes water to flow from the roots to the leaves?

a. translocation

b. transpiration

c. primary transport

d. secondary transport

e. transcytosis

16. What structures in your body are analogous to xylem and phloem?

a. liver and heart

b. arteries and veins

c. muscle and bone

d. gonads and mammary glands

e. nerves and endocrine glands

17. Root hairs:

a. increase the absorptive surface of the root.

b. are used to store food and nutrients.

c. provide support for the plant.

d. are the source for plant root growth.

e. have an as yet unknown function in plants.

18. Which of the following contains a diploid embryo?

i. seed

ii. fruit

iii. pollen

a. i only

b. ii only

c. iii only

d. i and ii

e. ii and iii

19. Guard cells control:

a. loading of sap into the phloem.

b. flow of fluid through the xylem.

c. opening and closing of the stomata.

d. absorption of water through root hairs.

e. opening and closing of flower buds.

20. Which of the following is *not* a part of the carpel?

a. stigma

b. style

c. anther

d. ovary

e. egg

21. Differential growth of a plant stem in response to a gravity gradient is known as:

a. gravitropism.

b. thigmotropism.

c. phototropism.

d. phytotropism.

e. photoperiodism.

22. A vine that climbs a tree to reach a sunny space in the tree canopy is using:

a. gravitropism.

b. thigmotropism.

c. phototropism.

d. phytotropism.

e. photoperiodism.

23. A seedling that bends toward the prevailing source of light from its position on the windowsill is experiencing:

a. gravitropism.

b. thigmotropism.

c. phototropism.

d. phytotropism.

e. photoperiodism.

24. Ragweed tends not to bloom far north of the equator because the nights are too short to trigger blooming during the optimal growing season. This is an example of:

a. gravitropism.

b. thigmotropism.

c. phototropism.

d. phytotropism.

e. photoperiodism.

25. You discover a new plant with some very odd adaptations that make it hard to classify using convential criteria. However, you do know that the plant drops its leaves during the dry season. At the very least then, you can classify this plant as:

a. a perennial.

b. an annual.

c. deciduous.

d. herbaceous.

e. dikaryotic.

WHAT'S IT ALL ABOUT?

Here's a question to help you pull together what you've learned so far using this text.

Question: We've all had the experience of watching a well-watered plant die. Given what you've learned about transport in this chapter, explain why too much water can be a bad thing for your houseplants.

What do I do now?

Remember the drill—decide what the question is asking you to do, collect your evidence from this chapter (and the others you've studied), and write!

CHAPTER 25 THE ANGIOSPERMS: FORM AND FUNCTION IN FLOWERING PLANTS

Basic Chapter Concepts

- Flowering plants vary in their anatomy and life cycles but share common cell types, transport systems, reproductive mechanisms, and growth patterns.
- Plant tissues are made up of three different kinds of cells—parenchyma, sclerenchyma, and collenchyma. These three cell types comprise the four tissue types (dermal, vascular, meristematic, and ground) of all angiosperms.
- All angiosperms reproduce sexually through an alternation of generations.

CHAPTER SUMMARY

25.1 Two Ways of Categorizing Flowering Plants
- Flowering plants can be classified by life span and by the number of embryonic leaves.
- Some plants complete their life cycles annually or biennially, and others live for many years.
- Important food crops have a single embryonic leaf and are classified as monocotyledons.
- About 75 percent of all flowering plants are dicotyledons—having two embryonic leaves.

25.2 There Are Three Fundamental Types of Plant Cells
- Parenchyma cells comprise most of the mature plant's living tissues. Parenchyma cells are multipotent—capable of serving as the precursors to sclerenchyma and collenchyma cells as well as dedifferentiating into embryonic tissues.
- Sclerenchyma cells provide structural support for the plant; their walls are strengthened by lignin as well as cellulose. Sclerenchyma cells are dead when mature.
- Collenchyma cells share the structural properties of sclerenchyma cells and the growth potential of parenchyma cells.

25.3 The Plant Body and Its Tissue Types
- Dermal tissue controls the plant's interactions with the outside world, regulating water and gas exchange, preventing infections, and secreting defense chemicals.
- Ground tissue, the most abundant tissue, is a prime site of photosynthesis.
- Vascular tissue consists of xylem and phloem, which regulate the movement of water and food, respectively.

25.4 How a Plant Grows: Apical Meristems Give Rise to the Entire Plant
- Meristematic tissue is the source of the vascular, ground, and dermal tissues because it is the only tissue capable of dividing, growing, and differentiating.
- Plants grow primarily at their tips (shoots and roots), and growth can continue indefinitely (is indeterminate). Woody plants can also grow laterally.

25.5 Secondary Growth Comes from a Thickening of Two Types of Tissue
- Herbaceous plants exhibit only primary growth—vertical growth from the tips of the roots and shoots. Woody plants, those that develop bark, exhibit both primary and secondary growth; secondary growth increases the plant's girth.

25.6 How the Plant's Vascular System Functions
- Flowering plants transport water from roots to tips through the xylem, a tissue that dies at maturity and supplies both transport and structural functions for the plant.

- Xylem transports water from root to tip, and the movement of water is driven by transpiration, which occurs when water is released from leaves. The energy for this process comes from the sun.
- Two types of cells, tracheids and vessel elements, make up the water-conducting elements of the xylem. Both of these cell types are nonliving in their mature, working state.
- Nutrients are transported through an adjacent system of cells called the phloem. The major product transported is sucrose, which is produced by photosynthesizing cells either in the leaves or stem.
- The main cell that actually transports fluid in the phloem is the sieve element, which lacks a nucleus at maturity. Its work is supported by another type of phloem cell, the companion cell, which loads the sucrose into the sieve element.
- Osmotic pressure, resulting from the difference in concentration between solutes outside plant cells and solutes inside the cell, powers the process. The plant tissue that is actively photosynthesizing and therefore producing sugar is called the source. The sugar is then transported to a sink, where it will be stored or immediately used.

25.7 Sexual Reproduction in Flowering Plants
- Plants produce spores that grow into a distinct generation, producing gametophyte plants that look different from the parental plant. However, many of the sporophyte generation stay associated with the parental plants, so it is hard to distinguish these plants from their parents.
- Gametophyte plants, the spore generation, produce haploid gametes (through meiosis) that fuse, producing a diploid zygote that will grow into a sporophyte plant.
- Flowers are the reproductive parts of the gametophyte generation, producing male gametes (pollen grains) and female gametes (eggs). Fusion of male and female gametes produces an embryo that becomes surrounded by fruit and a seed coat, both derived from the plant ovary. Sperm develop within pollen grains; eggs develop within the megaspore in the ovary.
- Fertilization results in production of a seed—a structure containing the developing embryo, its food supply, and a protective outer coat.

25.8 Embryo, Seed, and Fruit: The Developing Plant
- Once fertilization has occurred, a multistep process starts in which the ovule develops a seed coat that protects and will eventually provide a nutrient source for the developing embryo. The embryonic tissue develops two cotyledons, or embryonic leaves. As the seed germinates, it is these cotyledons that will emerge from the ground as a sprout.
- In some plants, the outer seed coat eventually thickens into fruit, some of which are soft and fleshy (e.g., apricot or apple), or others that are hard (e.g., corn kernel).
- The seed enclosed within the fruit can then be dispersed throughout the environment by various means, including animals, wind, and water. Seeds are very resilient and will then begin the growth cycle when environmental conditions are appropriate.

WORD ROOTS

angio- = blood or lymph vessels; **sperm-** = seed, germ (e.g., *angiosperms* are a group of plants that have flowers and produce seeds enclosed by tissues)

bi- = twice (e.g., a *biennial* is a flowering plant that requires two growing seasons to complete its life cycle)

hypo- = under, beneath; **-cotyl** = cotyledon (e.g., the *hypocotyl* is the part of the stem of an embryo plant that lies beneath the stalks of the seed leaves [cotyledons] and directly above the root)

trans- = through; **respire** = breathe (*transpiration* is evaporative water loss from a plant's aboveground parts, especially from stomata in the leaves)

KEY TERMS

angiosperm _____

_____

annual _____

_____

apical meristem _____

_____

bark _____

_____

biennial _____

bud _____

cellulose _____

collenchyma cell _____

companion cell _____

cork _____

cork cambium _____

cotyledon _____

dermal tissue _____

dicotyledon (dicot) _____

double fertilization _____

embryo sac _____

epicotyl _____

fruit _____

herbaceous plant _____

hypocotyl _____

intercalary meristem _____

lateral bud _____

megaspore mother cell _____

microspore mother cell _____

monocotyledon (monocot) _____

parenchyma cell _____

perennial _____

phloem _____

pollination _____

pressure-flow model _____

primary growth _____

root cap _____

sclerenchyma cell _____

secondary growth _____

secondary phloem _____

secondary xylem _____

seed coat _____

shoot apical meristem _____

sieve element _____

sieve tube _____

spore _____

terminal bud _____

tissue _____

tracheid _____

transpiration _____

vascular cambium _____

vessel element _____

wood _____

woody plant _____

_____

xylem _____

_____

xylem sap _____

_____

zone of cell division _____

_____

zone of differentiation _____

_____

zone of elongation _____

_____

FLASH CARDS

To use the flash cards, tear the page from the book and cut along the dotted lines. The key term appears on one side of the flash card, and its definition appears on the opposite side.

angiosperm	collenchyma cell
annual	companion cell
apical meristem	cork
bark	cork cambium
biennial	cotyledon
bud	dermal tissue
cellulose	dicotyledon (dicot)

the type of plant cell that provides support to allow the growing parts of the plant to stretch and elongate

a flowering seed plant whose seeds are enclosed within the tissue called fruit; angiosperms are the most dominant and diverse of the four principal types of plants; examples include roses, cacti, corn, and deciduous trees

in plants, cells that are closely associated with sieve elements in the fluid-conducting phloem tissue; the companion cells provide housekeeping needs of the sieve elements, which have lost their nuclei to provide room for faster conduction of phloem sap

a type of plant that goes through its entire life cycle—from germination of the seed through growth, flowering, and death—in one year

cells, dead in their mature state, that form the outermost covering of woody plants; these cells are infused with a waxy substance that protects the plant from drying out and from invaders

the group of plant cells at the tips of the roots and shoots that gives rise to all tissues in the plant

secondary meristematic tissue in woody plants that forms the outer living covering of woody plants; this cambium produces the cork cells that, in their dead, mature state, protect the outside of the trunk and branches

in woody plants, all the tissue layers outside the vascular cambium; from interior to exterior: the secondary phloem, phelloderm, cork cambium, and cork

an embryonic leaf; two major divisions in plants are the monocots, which have one embryonic leaf, and the dicots, which have two embryonic leaves

a type of plant that goes through its life cycle in about two years, flowering in the second year

the epidermis, or outer layer of cells of a plant; in addition to covering the plant, dermal tissue forms trichomes, such as root hairs

an undeveloped plant shoot, composed mostly of meristematic tissue

a type of plant that has two embryonic leaves within its seed; more than three-quarters of all flowering plants are dicotyledons

a complex carbohydrate that is the largest single component of plant cell walls; cellulose is dense and rigid and provides structure for much of the natural world; mammals cannot digest cellulose, so it serves as insoluble dietary fiber that helps move food through the digestive tract

double fertilization

lateral bud

embryo sac

megaspore mother cell

epicotyl

microspore mother cell

fruit

monocotyledon (monocot)

herbaceous plant

parenchyma cell

hypocotyl

perennial

intercalary meristem

phloem

an undeveloped plant shoot, located between the stem and a leaf, that may give rise to a branch or a flower or even take over the role of shoot apex if the apical meristem becomes damaged

---

in plants, the fusion of one sperm with the egg and another sperm with the central cell; the first of these fertilizations results in the zygote, which will develop into the embryo; the second produces the endosperm, which will provide food for the embryo; double fertilization occurs almost exclusively in flowering plants

---

a single diploid cell in the ovary of a flowering plant that undergoes meiosis, producing four megaspores, one of which develops into the mature female gametophyte plant

---

the mature female gametophyte plant consisting of seven cells produced by the megaspore; these cells include the egg and the central cell, which contains two nuclei

---

a diploid cell type, found within the anthers of flowering plants, that undergoes meiosis to produce the haploid microspores that develop into the mature male gametophyte plant

---

all tissue of an embryonic or seedling plant above the cotyledons; the epicotyl gives rise to the first true leaves

---

a type of plant that has one embryonic leaf within the seed; although comprising only one-quarter of all flowering plants, most important food plants are monocotyledons

---

the mature ovary of any flowering plant; many fruits protect the underlying seeds, and many attract animals that will eat the fruit and disperse the seeds

---

the most abundant type of cell in plants; these cells have thin cell walls, are usually alive at maturity, and serve numerous functions within the plant, including giving rise to the other plant cell types

---

a plant that never develops wood (secondary xylem), and therefore has relatively thin shoots; herbaceous plants do not undergo secondary growth (thickening)

---

a type of plant that lives for many years, such as trees, woody shrubs, and many grasses

---

all tissue of an embryonic or seedling plant below the cotyledons; the hypocotyl includes the plant's radicle, or early root structure, and is first to emerge from a seed when it starts to sprout

---

in vascular plants, the fluid-transporting tissue that conducts the food produced in photosynthesis, along with some hormones and other compounds

---

meristematic tissue found at the base of each growing node in grasses; these multiple growing regions permit fast growth and also allow grasses to keep growing even when their tops are removed by grazing or mowing

pollination	secondary xylem
pressure-flow model	seed coat
primary growth	shoot apical meristem
root cap	sieve element
sclerenchyma cell	sieve tube
secondary growth	spore
secondary phloem	terminal bud

xylem tissue produced to the inside of the vascular cambium in woody plants; this tissue, also called wood, is responsible for much of the thickening of woody plants

the transfer of pollen—by wind, animal, or other means—to a plant's female reproductive structure

the tough outer layer of a seed, derived from the integuments of the ovule, which protects the embryo from mechanical damage and water loss

the hypothesis that the force behind the movement of phloem sap in plants comes from the pressure produced in phloem when water moves into it by means of osmosis

the growing tip of plant shoots that produces the cells responsible for vertical growth of the plant

the type of growth in plants that occurs at the tips of their roots and shoots and mainly increases their length

in plants, a type of cell that conducts phloem sap

a collection of cells at the very tip of a root that protects the adjacent meristematic cells; the root cap secretes a lubricant to help the root move through the medium it is growing in

a tube, formed by a stack of sieve elements, that conducts phloem sap

a type of plant cell that is dead at maturity, has thick cell walls containing cellulose, and helps the plant return to its original shape after it has been deformed; these cells provide strong support for the plant

a reproductive cell that can develop into a new organism without fusing with another reproductive cell (the term *spore* also refers to a dormant, stress-resistant form of a bacterial or fungal cell)

the type of lateral growth in plants that thickens the plant; secondary growth occurs in woody plants, such as trees, and thickens their trunks and branches

another term for the shoot apex, or the main apical meristem of a plant; used especially when it is dormant

phloem tissue produced to the outside of the vascular cambium in woody plants as part of secondary growth, or thickening

tissue	**woody plant**
tracheid	**xylem**
transpiration	**xylem sap**
vascular cambium	**zone of cell division**
vessel element	**zone of differentiation**
wood	**zone of elongation**

a plant whose tissues include wood, or secondary xylem	an organized assemblage of similar cells that serves a common function; nervous, epithelial, and muscle tissue are some familiar examples
the tissue through which water and dissolved minerals flow in vascular plants	in plants, a slender, tapered cell of the xylem that has perforations (bordered pits) that match up with adjacent cells to allow water to move between them
the water and dissolved minerals that are conducted through xylem in plants	the process by which plants lose water when water vapor leaves the plant through open stomata; more than 90 percent of the water that enters a plant evaporates into the atmosphere via transpiration
in plants, the area just behind the apical meristem, where more cells are produced; cells then move on to the zones of elongation and differentiation	the thin layer of tissue in woody plants between the primary xylem and phloem tissue layers that generates the secondary xylem and phloem tissues; this meristematic tissue brings about secondary growth in woody plants
in plants, the area just behind the zone of elongation, near the apical meristem, where primary tissue types finish taking shape	one of two types of cells that make up xylem tissue, which transports water and dissolved minerals through plants; at maturity, vessel elements are dead and empty of cellular contents, thus facilitating the rapid conduction of water
in plants, the area between the zones of cell division and differentiation, near the apical meristem, where most primary plant growth occurs through the elongation of cells	secondary xylem in a plant; this material in woody plants is responsible for most of their thickening

SELF TEST

Once you have finished studying this chapter, close your books, grab a pencil, and spend the next 15 to 20 minutes completing this practice test.

Compare and Contrast

For each of the following paired terms, write a sentence of comparison ("Both") and a sentence of contrast ("However,").

pollen/endosperm
monocot/dicot
primary/secondary growth
cork cambium/vascular cambium
epicotyl/hypocotyl

Short Answer

1. Describe the two ways of classifying plants. Which method would the home gardener most likely use?

2. What plant cell type is most analogous to an animal stem cell? How could this be demonstrated?

3. You have a maple leaf before you. Describe where the various plant tissues can occur in this leaf.

4. If you accidentally cut off the top of your favorite plant when trimming it, will it continue to grow? Give a reason for your answer.

5. Why does a tree die if you remove a ring of its bark?

6. Is water pushed or pulled through a plant? What causes water to flow from the roots to the leaves?

7. Describe the events that have to occur before fertilization occurs in a flowering plant.

8. What are the differences among seed, fruit, and pollen?

Multiple Choice

Circle the letter that best answers the question.

1. Secondary xylem and secondary phloem form which of the following?
   a. epididymis
   b. bark
   c. gametophytes
   d. chloroplasts
   e. none of the above

2. What type of cell forms the bulk of a herbaceous plant and carries out a variety of functions in that plant?
   a. collenchyma
   b. parenchyma
   c. meristems
   d. vascular bundles
   e. sclerenchyma

3. What happens in the zone of differentiation?
   a. growth
   b. cell lengthening
   c. cell specialization
   d. phelloderm production
   e. none of the above

4. What type of tissue makes up the bulk of a tree trunk?
   a. primary xylem
   b. primary phloem
   c. secondary xylem
   d. secondary phloem
   e. cork

5. Which cells are dead at maturity?
   a. tracheids
   b. companion cells
   c. vessel collenchyma
   d. sclerenchyma
   e. parenchyma

6. What is the ploidy of endosperm?
   a. haploid
   b. diploid
   c. triploid
   d. aneuploid
   e. depends on the plant

7. A mature pollen grain consists of all of the following *except:*
   a. an outer coat of ground tissue.
   b. two sperm cells.
   c. two egg cells.
   d. one tube cell.
   e. None of the above is part of a mature pollen grain.

8. How do sperm reach the ovule?
   a. by entering through leaf stomata
   b. by entering through the root cortex
   c. by the pollen tube
   d. through ingestion by an animal pollinator
   e. by chloroplastic immigration

9. If the surrounding air is saturated with water (100 percent humidity), what would happen to xylem flow?
   a. It would move faster.
   b. It would reverse direction.
   c. It would slow or stop.
   d. Water content of air has nothing to do with xylem flow.
   e. Plant stomata would open.

10. After a 10-year absence, you return to the house where you grew up. Outside in the yard, you find that your old tree house is still in good shape. You decide to climb the ladder nailed into the side of the

oak tree one last time to reminisce. You notice that after 10 years of tree growth, the steps of the ladder:
  a. have moved upward (away from the ground).
  b. have moved downward (toward the ground).
  c. have not changed in height.
  d. have increased in width.
  e. have decreased in width.

11. Most of the water moving into a leaf is lost through:
  a. osmosis.
  b. transpiration.
  c. pressure-flow forces.
  d. translocation.
  e. all of the above

12. Which of the following functions below are *not* a function of parenchyma cells?
  a. support
  b. heal wounds
  c. store food
  d. conduct photosynthesis
  e. regenerate lost parts

13. How do cells in a meristem differ from other plant cells?
  a. They continue to divide.
  b. They photosynthesize at a faster rate.
  c. They are growing.
  d. They are differentiating into other cells.
  e. They store food.

14. Celery stalks are fibrous and stringy. The stalks must contain a lot of:
  a. parenchyma cells.
  b. phloem.
  c. meristematic tissue.
  d. sclerenchyma cells.
  e. epidermal cells.

15. Which of the following is *not* a characteristic of dicots?
  a. two seed leaves
  b. parts of flowers in fours or fives
  c. taproot
  d. vascular bundles arranged in a circle
  e. parallel venation in leaves

16. Guard cells:
  a. control the rate of transpiration.
  b. push water upward in a plant stem.
  c. protect the plant roots from infection.
  d. control water and solute intake by roots.
  e. protect bacteria that are located in nodules on the roots.

17. If you place a plastic bag over a plant overnight, in the morning the plastic will be fogged with water drops. This is due to:
  a. root pressure.
  b. adhesion.
  c. photosynthesis.
  d. pressure flow.
  e. transpiration.

18. Sugars are carried throughout the plant in which tissue?
  a. cortex
  b. cambium
  c. xylem
  d. phloem
  e. parenchyma

19. What are the three basic tissue types found in plants?
  a. xylem, phloem, and sieve tube
  b. parenchyma, sclerenchyma, and collenchyma
  c. ground, vascular, and epidermal
  d. meristematic, wood, and parenchyma
  e. cork, cork cambium, and wood

20. Wood consists of annual layers of:
  a. phloem.
  b. fibers.
  c. bark.
  d. parenchyma.
  e. xylem.

21. Which of the following is not a fruit?
  a. tomato
  b. peanut
  c. pumpkin
  d. potato
  e. plum

22. The endosperm of a plant is:
  a. found in a seed.
  b. found in a stem.
  c. found in a shoot.
  d. found in the leaves.
  e. found only in gymnosperms.

23. The sink region in the pressure-flow hypothesis of phloem transport is most often the:
  a. growing leaves.
  b. fruit.
  c. seeds.
  d. roots.
  e. all of the above

24. Cells in a meristematic region of a plant:
  a. are very large.
  b. have thin cell walls.
  c. are very small.
  d. are usually green.
  e. do not contain a nucleus.

25. You are performing a chemical analysis of xylem sap. You probably would not expect to find much of which of the following?
  a. nitrogen
  b. sugar
  c. phosphorus
  d. water
  e. potassium

WHAT'S IT ALL ABOUT?

Here's a question to help you pull together what you've learned so far using this text.

Question: Angiosperms are the most successful of the higher plants, both in terms of numbers and diversity. Thinking back over concepts included in this chapter, discuss five ways that flowering plants are critical to the biosphere and our survival.

What do I do now?

Remember the drill—decide what the question is asking you to do, collect your evidence from this chapter (and the others you've studied), and write!

CHAPTER 26 INTRODUCTION TO HUMAN ANATOMY AND PHYSIOLOGY: THE INTEGUMENTARY, SKELETAL, AND MUSCULAR SYSTEMS

Basic Chapter Concepts

- Human bodies, like those of other mammals, are characterized by the presence of a central body cavity, an internal skeleton, and a stable body temperature.
- Humans have organs composed of specialized tissues. These organs function together, creating systems that carry out all the essential processes of the body—providing structural support, protecting the body against the environment, transporting nutrients in and wastes out, interacting with the environment, and maintaining a stable internal environment.

CHAPTER SUMMARY

26.1 The Disciplines of Anatomy and Physiology

- Humans, like all multicellular animals, require systems to deliver nutrients and remove wastes from the cells that are not in direct contact with the environment.
- These systems, composed of organs, reside within a central body cavity (coelom). The shape of the body is maintained by an internal skeleton, which also protects the organs. As the organ systems carry out their metabolic functions, heat is generated. Thus, humans are able to maintain a stable internal temperature.
- The field of biology that describes the structures comprising a body is called anatomy; physiology examines the functions of these structures.

26.2 How Does the Body Regulate Itself?

- The central concept of physiology is homeostasis, that is, the maintenance of body parameters within certain limits.
- Homeostasis is maintained almost entirely by negative feedback, in which a series of processes generates a response that reduces or destroys the original stimulus. This is similar to a thermostat regulating the heating system in a house or apartment.

26.3 Levels of Physical Organization

- Four different types of tissues exist in the human body: epithelial, connective, muscle, and nervous.
- Organs are collections of tissues that cooperate to perform a single function. Multiple types of tissues may be found in a single organ, so that different steps of the metabolic process can be carried out in the organ.

26.4 The Human Body Has Four Basic Tissue Types

- Each of the tissue types in the body has characteristic features.
- Epithelial tissues form a border or barriers between one medium and another. Examples are the skin (integument) and the gut lining.
- Connective tissues are quite diverse but are united by a typical structure: cells embedded in a matrix.
- Nervous tissue is made up of neurons and their support (glial) cells.
- Muscle tissue can be subdivided into three groups: cardiac, skeletal and smooth.

26.5 Organs Are Made of Several Kinds of Tissues

- Organs are a composite of different types of tissues. For example, the skin has both epithelial and connective tissue.

26.6   Organs and Tissues Make Up Organ Systems

- Organs work in concert to carry out bodily functions. There are 11 organ systems in the human body: integumentary, skeletal, muscular, nervous, endocrine, lymphatic, cardiovascular, respiratory, digestive, urinary, and reproductive.

26.7   The Integumentary System: Skin and Its Accessories

- Covering the body and maintaining body temperature are the functions of the integumentary system, which consists of the skin and its associated structures and glands.
- Our skin is the largest organ of our bodies. It is made of two layers, the dermis and the epidermis, and serves as the first line of defense, protecting the internal organs from the environment. Hair, sweat and sebaceous glands, and nails are skin-associated structures that help us regulate our body temperature and provide protection.

26.8   The Skeletal System

- The skeletal system counters the effects of gravity. Bone tissue is mostly connective tissue strengthened by large deposits of calcium compounds. Within the bone exists a loose connective tissue—marrow—that makes blood tissue. Bones connect at the joints, highly mobile structures that hold bones together without the bones touching, so that the body has flexibility.

26.9   The Muscular System

- Muscle produces movement by contraction and relaxation. During a contraction, muscle fibers (collections of muscle-tissue cells) slide past each other, bringing the attached bones closer together. When the muscle relaxes, the fibers slide apart, separating the bones.

WORD ROOTS

endo- = within (e.g., *endo*crine glands release their products within [or to the interior] of the body)

exo- = outside (e.g., *exo*crine glands release their products to the gut or the exterior of the body)

KEY TERMS

appendicular skeleton _____

axial skeleton _____

bone _____

cardiac muscle _____

cardiovascular system _____

cartilage _____

compact bone _____

connective tissue _____

dermis _____

digestive system _____

endocrine gland _____

endocrine system _____

epidermis _____

epithelial tissue _____

exocrine gland _____

gland _____

homeostasis _____

hormone _____

immune system _____

integumentary system _____

keratin _____

_____

ligaments _____

_____

lymphatic network _____

_____

muscle fiber _____

_____

muscle tissue _____

_____

muscular system _____

_____

negative feedback _____

_____

nervous system _____

_____

nervous tissue _____

_____

organ _____

_____

organ system _____

_____

osteoblast _____

osteoclast _____

osteocyte _____

_____

red marrow _____

_____

reproductive system _____

_____

respiratory system _____

_____

sarcomere _____

_____

sebaceous glands _____

_____

skeletal muscle _____

_____

skeletal system _____

_____

skin _____

_____

smooth muscle _____

_____

spongy bone _____

_____

sweat gland _____

_____

tendons _____

_____

tissue _____

_____

urinary system _____

_____

yellow marrow _____

_____

FLASH CARDS

To use the flash cards, tear the page from the book and cut along the dotted lines. The key term appears on one side of the flash card, and its definition appears on the opposite side.

appendicular skeleton	connective tissue
axial skeleton	dermis
bone	digestive system
cardiac muscle	endocrine gland
cardiovascular system	endocrine system
cartilage	epidermis
compact bone	epithelial tissue

a tissue, active in the support and protection of other tissues, whose cells are surrounded by a material that they have secreted; in humans, one of the four principal types of tissue

the division of the skeletal system consisting of the bones of the paired appendages, including the pelvic and pectoral girdles to which they are attached

in certain animals, the thick layer of the skin—composed mostly of connective tissue—that underlies, nourishes, and supports the epidermis

the division of the skeletal system that forms the central column, including the skull, vertebral column, and rib cage

the organ system that transports food into the body, secretes digestive enzymes that help break down food to allow it to be absorbed by the body, and excretes waste products; this system consists of the esophagus, stomach, and large and small intestines, plus the accessory glands that produce the enzymes along the way

a connective tissue that provides support and structure to the body and that often is the site of fat storage and blood cell production

a gland that releases its materials directly into surrounding tissues or into the bloodstream, without using ducts; many hormones are produced by endocrine glands

the type of striated muscle tissue that forms the muscles of the heart

the organ system that sends signals throughout the body through use of the chemical messengers called hormones

a fluid transport system of the body, consisting of the heart, all the blood vessels in the body, the blood that flows through these vessels, and the bone marrow tissue in which red blood cells are formed

the outermost layer of skin in animals or the outermost cell layer in plants

a connective tissue that serves as padding in most joints, forms the human larynx (voice box) and trachea (windpipe), and links each rib to the breastbone

a tissue that covers surfaces exposed to an external environment; in humans, skin is an epithelial tissue, as is the lining of the digestive tract

dense bone that forms the outer portion of bones, structured as a set of parallel osteons and their associated nerves and blood vessels

exocrine gland

gland

homeostasis

hormone

immune system

integumentary system

keratin

ligaments

lymphatic network

muscle fiber

muscle tissue

muscular system

negative feedback

nervous system

in anatomy, a connective tissue that links one bone to another

a gland that secretes its materials through ducts (tubes); for example, sweat glands conduct perspiration through ducts to the skin

in humans, the transport network that collects interstitial fluid, transports it as lymph through lymphatic vessels, checks the fluid for infection, and delivers the fluid to blood vessels

an organ or group of cells that secretes one or more substances

a single skeletal muscle cell; called a fiber because of its extreme length relative to most cells

the maintenance of a relatively stable internal environment in living things

tissue that has the ability to contract; in humans, one of the four principal types of tissue

a substance that, when released in one part of an organism, goes on to prompt physiological activity in another part of the organism; both plants and animals have hormones

the organ system composed of all the skeletal muscles of the body, which is to say all muscles that are under voluntary control

the collection of cells and proteins that, in mammals, function together to kill or neutralize invading microorganisms

a system of control in which the product of a process reduces the activity that led to the product

the organ system that protects the body from the external environment and assists in regulating body temperature; this system consists of the skin and associated structures, such as glands, hair, and nails

the organ system that monitors an animal's internal and external environment, integrates the sensory information received, and coordinates the animal's responses; this system consists of all the body's neurons, plus the supporting neuroglia cells, plus the sensory organs

a flexible, water-resistant protein, abundant in the outer layers of skin, that also makes up hair and fingernails

nervous tissue

reproductive system

organ

respiratory system

organ system

sarcomere

osteoblast

sebaceous glands

osteoclast

skeletal muscle

osteocyte

skeletal system

red marrow

skin

in humans, the organ system that develops gametes and delivers them to a location where they can fuse with other gametes to produce a new individual

a tissue specialized for the rapid conduction of electrical impulses; in humans, one of the four principal tissue types

the organ system that brings oxygen into the body and expels carbon dioxide from the body; in humans, this system includes the lungs and passageways that carry air to the lungs

a highly organized unit within an organism, performing one or more functions, that is formed of several kinds of tissue; kidneys, heart, lungs, and liver are all familiar examples of organs in humans

the functional unit of a striated muscle that contracts when thin filaments slide past thick filaments; the sarcomeres shorten, thus contracting the whole muscle

a group of interrelated organs and tissues that serve a particular set of functions in the body; for example, the digestive system consists of mouth, stomach, and intestines and functions in digesting food and eliminating waste

a type of gland in the skin that produces a waxy, oily secretion (sebum) that lubricates the hair shaft and inhibits bacterial growth in the surrounding area

an immature bone cell that secretes organic material that becomes bone matrix, thus producing new bone

in humans, muscle that is attached to bone, that is under conscious control, and that microscopically has a striped or "striated" appearance owing to the parallel orientation of the long, fibrous units that make it up

a type of bone cell that dissolves bone matrix, thus liberating the minerals stored in it

the human organ system that forms an internal supporting framework for the body and protects delicate tissues and organs; this system consists of all the bones and cartilages in the body and the connective tissues and ligaments that connect the bones at the joints

mature bone cell that maintains the structure and density of bone by continually recycling calcium compounds around itself

in humans, an organ consisting of two tissue layers, an outer epidermis and inner dermis, and covering the outside of the body; the skin protects the body and receives signals from the environment

a tissue, found in cavities of bones in the human body, within which all of the adult body's blood cells are produced

smooth muscle

tissue

spongy bone

urinary system

sweat gland

yellow marrow

tendons

an organized assemblage of similar cells that serves a common function; nervous, epithelial, and muscle tissue are some familiar examples

in humans, muscle that is not under voluntary control and that lacks a striated appearance; smooth muscle is responsible for contractions of the uterus, digestive tract, blood vessels, and passageways of the lungs

the organ system that eliminates waste products from the blood through formation of urine; in humans, this system consists of the kidneys, where the urine is formed; and the ureters, urinary bladder, and urethra, which transport the urine from the kidneys to the outside of the body

type of bone that is porous and less dense than compact bone; spongy bone fills the expanded ends of long bones

in human beings, a tissue largely made up of energy-storing fat cells found in the marrow cavity of long bones

in humans, a type of duct-containing (exocrine) gland that produces perspiration

the connective tissue that attaches a skeletal muscle to a bone

SELF TEST

After you have finished studying this chapter, close your books, grab a pencil, and spend the next 15 to 20 minutes completing this practice test.

Compare and Contrast

For each of the following paired terms, write a sentence of comparison ("Both") and a sentence of contrast ("However").

cardiac/skeletal muscle
exocrine/endocrine glands
basement membrane/ground substance
osteoclast/osteoblast
dermis/epidermis

Short Answer

1. How are connective tissues defined?

2. Explain the functions of the three types of muscle tissue.

3. What is the function of the exocrine glands? Give an example.

4. Explain the difference in function between red and yellow bone marrow.

5. Give a specific example of an organ that contains several different types of tissues.

6. How is negative feedback in the human body similar to the heating system in your home?

7. What is the function of ligaments and tendons at articulations?

8. Compare and contrast sebaceous and sweat glands in terms of function.

9. What is homeostasis? Why is it important to understanding physiology?

Multiple Choice

Circle the letter that best answers the question.

1. There is/are _____ main tissue type(s) in animals.
   a. one
   b. two
   c. three
   d. four
   e. five

2. Blood is a type of:
   a. elastic tissue.
   b. epithelium.
   c. muscle.
   d. connective tissue.
   e. nervous tissue.

3. What mineral helps to harden bone?
   a. magnesium
   b. carbonate
   c. calcium
   d. zinc
   e. iron

4. Sweat is produced by:
   a. apocrine glands.
   b. sebaceous glands.
   c. hair follicles.
   d. keratin.
   e. a and b

5. Which of the following is *not* a type of connective tissue?
   a. blood
   b. bone
   c. ligaments
   d. skin
   e. tendons

6. Which of the following is *not* a characteristic of mammals?
   a. internal body cavity
   b. temperature regulation
   c. internal skeleton
   d. single-celled organism
   e. hair

7. Connective tissue:
   a. includes muscle cells.
   b. is the most diverse of the tissue types.
   c. is found only among plants.
   d. is classified as squamous, cuboidal, and stratified.
   e. all of the above

8. The function of the lymph system is to:
   a. produce hormones.
   b. produce skin cells.
   c. protect the body from invasion.
   d. transmit messages.
   e. make blood cells.

9. The cardiovascular, respiratory, digestive, and urinary systems all share which of the following characteristics?
   a. They consist solely of connective tissue.
   b. They have no organs.
   c. They have structures within the skull.
   d. They transport and exchange materials with the environment.
   e. all of the above

10. Every organ system:
    a. is made up of several different types of tissues.
    b. can survive independently outside the body.
    c. is classified as a -cyte, -blast, or –clast.
    d. has cells suspended in liquid material.
    e. has pipes and pedals.

11. Your friend Serafina had a minor accident last week while rock climbing. Since then, she feels a sharp pain whenever she tries to use her left index finger. She thinks she may have broken a bone. You suggest that she consult a doctor who specializes in:
   a. muscle tissue.
   b. epithelial tissue.
   c. connective tissue.
   d. nervous tissue.
   e. facial tissue.

12. Scleroderma ("hard skin") is a disease that affects epithelial tissue. Aside from the skin, what other structures would you predict to be affected?
   a. lining of the stomach
   b. lining of the lungs
   c. heart muscle
   d. both a and b
   e. both b and c

13. Bone is:
   a. comprised of cardiac muscle.
   b. comprised of stratified squamous tissue.
   c. a form of connective tissue.
   d. involved in message transfer and storage.
   e. helps to form linings and coverings.

14. Nervous tissue is:
   a. comprised of cardiac muscle.
   b. comprised of stratified squamous tissue.
   c. a form of connective tissue.
   d. involved in message transfer and storage.
   e. helps to form linings and coverings.

15. The heart is:
   a. part of cardiac muscle.
   b. comprised of stratified squamous tissue.
   c. a form of connective tissue.
   d. involved in message transfer and storage.
   e. helps to form linings and coverings.

16. Epithelium could be classified as:
   a. cardiac muscle.
   b. stratified squamous tissue.
   c. connective tissue.
   d. message transfer and storage.
   e. smooth or glial.

17. Cartilage is:
   a. cardiac muscle.
   b. epithelium.
   c. connective tissue.
   d. smooth muscle.
   e. nervous tissue.

18. Which of the following would be classified as part of the study of anatomy?
   a. the shape and placement of your left little toe
   b. the conversion of doughnuts into energy sources
   c. the transfer of messages from the brain to the muscles

   d. the regulation of urine formation in the kidney
   e. the formation of gametes (eggs and sperm)

19. Put the following in order from simplest to most complex:
   a. organs; tissues; organ systems; cells; molecules
   b. organs; organ systems; cells; tissues; molecules
   c. tissues; organ; organ systems; molecules; cells
   d. cells; molecules; tissues; organ; organ systems
   e. molecules; cells; tissues; organ; organ systems

20. Of the terms listed in question 19, which refers to the smallest unit of life?
   a. organ system
   b. cell
   c. tissue
   d. molecule
   e. organ

21. Muscles contract by the movement of actin and myosin. Place the following steps in the correct order of the contraction cycle:
   i. Myosin pulls actin filament to the middle of the sarcomere.
   ii. Myosin heads attach to the actin filament.
   iii. Myosin heads detach from the actin filament.
      a. i, ii, iii
      b. i, iii, ii
      c. ii, i, iii
      d. ii, iii, i
      e. iii, ii, i

22. A new deodorant is designed to inhibit body odor. Which of the following should it be designed to interact with?
   a. merocrine glands
   b. apocrine glands
   c. terocrine glands
   d. endocrine glands
   e. mesocrine glands

23. Which three cell types are found in bone?
   a. osteocytes
   b. osteoblasts
   c. osteoclasts
   d. a and b
   e. a, b, and c

24. Which organ system forms the internal fluid transport system?
   a. lymphatic system
   b. cardiovascular system
   c. urinary system
   d. integumentary system
   e. nervous system

25. Which type of tissue contains the stored excess energy from your chocolate binges?
   a. connective
   b. nervous
   c. smooth muscle
   d. skeletal muscle
   e. epithelial

Use the figure below to answer questions 26–29.

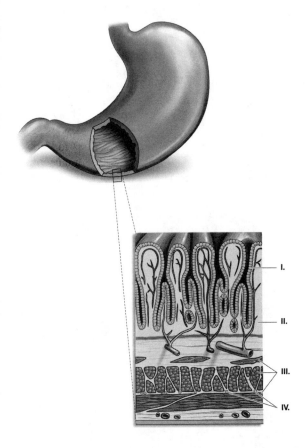

c. III
d. IV
e. none of the above

29. Which type includes blood and adipose tissue as well?
   a. I
   b. II
   c. III
   d. IV
   e. All of these include blood and adipose tissue as well.

Use the figure below to match the correct term with its corresponding letter for questions 30–36.

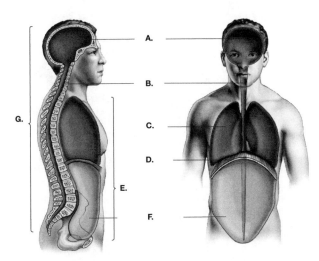

26. Which of these types of tissue also includes the heart?
   a. I
   b. II
   c. III
   d. IV
   e. both I and II

27. Which type is nervous tissue?
   a. I
   b. II
   c. III
   d. IV
   e. both III and IV

28. Which type also forms the exchange surface of the lung?
   a. I
   b. II

30. _____ spinal cavity
31. _____ ventral cavity
32. _____ cranial cavity
33. _____ dorsal cavity
34. _____ abdominopelvic cavity
35. _____ thoracic cavity
36. _____ diaphragm

WHAT'S IT ALL ABOUT?

Here's a question to help you pull together what you've learned so far using this text..

Question: In the previous unit, we learned about the basic structure and function of plant tissues. Can we identify similar types of tissues in animals? Do they have similar functions?

What do I do now?

Remember the drill—decide what the question is asking you to do, collect your evidence from this chapter (and the others you've studied), and write!

CHAPTER 27   COMMUNICATION AND CONTROL: THE NERVOUS AND ENDOCRINE SYSTEMS

Basic Chapter Concepts

- The nervous system processes information received from both inside and outside the body through a network of sensory neurons. Information is processed either by the central nervous system (CNS) or directly by the peripheral nervous system (PNS). The autonomic and somatic divisions of the PNS carry information directing a response to the appropriate muscles and organs.
- The endocrine system produces the hormones, the chemical messengers responsible for maintaining internal homeostasis in the face of a constantly changing external environment.

CHAPTER SUMMARY

27.1   Structure of the Nervous System
- The nervous system is composed of the brain and spinal column (CNS) and the neurons that transduce information to and from the responding organs (PNS). Some responses to nerve impulses are under voluntary control (somatic), and some are nonvoluntary (autonomic).

27.2   Cells of the Nervous System
- Primary cells of the nervous system are neurons; sensory neurons deliver information from the sensory organs, motor neurons deliver information to muscles, and interneurons carry information between the two. The protective neuroglia cells surround neurons in the CNS.

27.3   How Nervous System Communication Works
- Information travels in the nervous system as a change in the distribution of charge across the membrane; this change in membrane potential is called an action potential because it travels unidirectionally from a neuron cell body to the end of the axon.
- Neurons do not touch, so action potentials must use neurotransmitter molecules to travel across the small space (synapse) separating one axon from the cell body of another neuron.

27.4   The Spinal Cord
- The spinal cord serves as a conduit for messages between the PNS and the brain. Some incoming messages to the spinal cord elicit immediate, efferent responses to the muscles—these are the reflexes, the simplest response of the autonomic nervous system.

27.5   The Autonomic Nervous System
- Involuntary responses may either conserve energy (parasympathetic response) or stimulate the demand for energy (sympathetic response). Both divisions of the autonomic nervous system innervate organs, so organ function represents a balance between the two messages.

27.6   The Human Brain
- The human brain consists of six major regions: the cerebrum, thalamus and hypothalamus, midbrain, pons, cerebellum, and medulla oblongata. The cerebrum is the site of higher thinking.
- The brain stem consists of the midbrain, pons, and medulla oblongata and controls involuntary body activities. The thalamus receives sensory perceptions, which are sent on to the cerebrum. The hypothalamus maintains homeostasis.

27.7   The Nervous System in Action: Our Senses
- The senses relay information to the brain. Each sensory organ performs two functions: responding to stimuli and transforming the response into an electrical signal that is sent to the brain.

27.8 Our Sense of Touch

- Touch is based on sensory receptors that can respond to light or heavy pressure and various types of skin contact. The sensory information is then transformed into an electrical signal and sent to the brain via sensory neurons.

27.9 Our Sense of Smell

- The sense of smell depends on sensory neurons with dendritic extensions into the nasal passages. Molecules with identifiable smells bind to these dendrites and trigger an electrical signal that is sent to the brain.

27.10 Our Sense of Taste

- The sense of taste is conveyed through the taste buds located on the tongue. Taste buds respond to sweet, sour, salty, and bitter tastes. Sensory neurons receive input from the taste buds, which is then sent as electrical signals to the brain.

27.11 Our Sense of Hearing

- The organ of hearing is the ear, which contains sensitive hair cells that respond to waves of compression against the eardrum. The opening and closing of ion channels in the cochlea results in electrical signals to the brain.

27.12 Our Sense of Vision

- Vision is based on three components: (1) gathering and focusing of light; (2) conversion of light signals into electrical signals; and (3) an appropriate response to the visual information.
- The central organ of vision is the eye, which has three vital regions: (1) a light-sensitive region at the back of the eye, called the retina; (2) photoreceptor cells in the retina, some of which can respond to dim light (rods), and others of which respond best in bright light (cones); and (3) the optic nerve, which relays electrical signals to the brain.

27.13 The Endocrine System

- Organs that synthesize chemical messengers exert their effect on some other tissue via specific receptors, which can be bound to the membrane or inside the cell.

27.14 Types of Hormones

- Hormones may be amino acid-based, peptides, or steroid-based. Amino acid-based and peptide hormones must be detected by receptors on the surface of the target cell, whereas lipid-based hormones can diffuse through cell membranes.

27.15 How Is Hormone Secretion Controlled?

- Most hormone secretion is controlled by negative feedback, in which the release of the hormone eventually results in the suppression of further release of the hormone. Many negative feedback loops contribute to maintaining stability (homeostasis) within the internal environment of an organism.
- The hypothalamus is central to the control of hormonal secretion. The hypothalamus, via the posterior pituitary, produces two hormones: adrenalin and noradrenalin. Through the nervous system, the hypothalamus also controls the release of six additional hormones from the anterior pituitary.

27.16 Hormones in Action: Four Examples

- Levels of blood glucose are controlled through two important hormones secreted by the pancreas. Insulin reduces levels of blood glucose, whereas glucagon increases levels of blood glucose.
- Oxytocin, released by posterior pituitary, is important in promoting contractions during childbirth and milk release from nursing mothers.
- Cortisol, released by the adrenal glands, is known as the stress hormone because it increases energy levels in times of heightened stress. However, continual release of cortisol can have negative effects.

WORD ROOTS

homo- = one and the same; **stasis** = stable state (e.g., *homeostasis* is the maintenance of a relatively stable internal physiological environment in an organism)

photo- = light; **receptor** = receiver (e.g., a *photoreceptor* is a light-sensitive sensory cell of invertebrates and vertebrates)

soma- = the body of an organism (e.g., the *somatic* nervous system consists of the nerves that connect the vertebrate central nervous system and skeletal muscles)

KEY TERMS

action potential _____

afferent division _____

amino acid-based hormones _____

anterior pituitary _____

autonomic nervous system _____

axon _____

central nervous system (CNS) _____

cerebral cortex _____

cerebrospinal fluid _____

cerebrum _____

cochlea _____

cones _____

dendrites _____

efferent division _____

endocrine gland _____

ganglion _____

glucagon _____

homeostasis _____

hormones _____

hypothalamus _____

insulin _____

interneuron _____

membrane potential _____

motor neuron _____

myelin _____

nerve _____

neurotransmitter _____

olfaction _____

parasympathetic division _____

peptide hormones _____

peripheral nervous system (PNS) _____

photoreceptor _____

posterior pituitary _____

reflex _____

retina _____

rods _____

sensory neuron _____

somatic nervous system _____

steroid hormones _____

sympathetic division _____

synapse _____

synaptic cleft _____

target cells _____

thalamus _____

FLASH CARDS

To use the flash cards, tear the page from the book and cut along the dotted lines. The key term appears on one side of the flash card, and its definition appears on the opposite side.

action potential	central nervous system (CNS)
afferent division	cerebral cortex
amino acid-based hormones	cerebrospinal fluid
anterior pituitary	cerebrum
autonomic nervous system	cochlea
axon	cones

the portion of the nervous system consisting of the brain and spinal cord

a temporary reversal of neuronal cell membrane potential at one location that results in a conducted nerve impulse down an axon

thin outer covering of the region of the brain known as the cerebrum; responsible for the highest human thinking and processing

the division of the peripheral nervous system that carries sensory information toward the central nervous system, having gathered information about the body or environment

a fluid that circulates in both the brain and spinal cord, supplying nutrients, hormones, and immune system cells and providing the nervous tissue with protection against jarring injury

hormones that are derived from a single amino acid; one of three principal classes of hormones, the other two being peptide hormones and steroid hormones

the largest region of the human brain, responsible for much of the human capacity for higher mental functioning

an endocrine gland that releases two hormones that work directly on target cells and four other hormones that regulate the production of hormones by other endocrine glands

the coiled, fluid-filled, membranous portion of the inner ear, in which vibrations are transformed into the nervous system signals perceived as sound

that portion of the peripheral nervous system's efferent division that provides involuntary regulation of smooth muscle, cardiac muscle, and glands

in human vision, photoreceptors that respond best to bright light and that provide color vision

a single, large extension of the cell body of a neuron that carries signals away from the cell body toward other cells

dendrites

hormone

efferent division

hypothalamus

endocrine gland

insulin

ganglion

interneuron

glucagon

membrane potential

homeostasis

motor neuron

a substance that, when released in one part of an organism, goes on to prompt physiological activity in another part of the organism; both plants and animals have hormones

extensions of a neuron that carry signals toward the neuronal cell body

portion of the brain important in drives and in the maintenance of homeostasis; the latter capacity reflects the fact that the hypothalamus exerts control over a good deal of the body's hormonal release

the division of the peripheral nervous system that carries motor commands from the central nervous system (CNS) toward the effectors (muscles and glands)

a hormone, secreted by cells in the pancreas, that brings about a decrease in blood levels of glucose

a gland that releases its materials directly into surrounding tissues or into the bloodstream, without using ducts; many hormones are produced by endocrine glands

a type of neuron, located only within the brain or spinal cord, that connects other neurons; these neurons are responsible for the analysis of sensory inputs and the coordination of motor commands

any collection of nerve-cell bodies in the peripheral nervous system

the electrical charge difference that exists from one side of a neuron's plasma membrane to the other

a hormone, secreted by cells in the pancreas, that brings about an increase in blood levels of glucose

a neuron that sends instructions from the central nervous system (CNS) to such structures as muscles or glands; given the direction of this transmission, motor neurons are efferent neurons

the maintenance of a relatively stable internal environment in living things

myelin	**peptide hormones**
nerve	**peripheral nervous system (PNS)**
neurotransmitter	**photoreceptor**
olfaction	**posterior pituitary**
parasympathetic division	**reflex**

hormones composed of chains of amino acids, with these chains ranging from small polypeptides to proteins composed of hundreds of amino acids; one of three principal classes of hormones, the other two being amino acid-based hormones and steroid hormones

---

membranous covering of some neuronal axons, provided by glial cells, that allows faster nerve signal transmission through these axons

---

the part of the nervous system that includes all of the neural tissue outside the central nervous system (brain and spinal cord); the PNS brings information to and carries it from the central nervous system; it also provides voluntary control of the skeletal muscles and involuntary control of the smooth muscles, cardiac muscles, and glands

---

a bundle of axons in the peripheral nervous system that transmits information to or from the central nervous system

---

sensory receptor cell for vision, located in the retina, that transforms light into nervous system signals; photoreceptors come in two varieties: rods, which function in low-light situations but provide only black-and-white vision; and cones, which respond best to bright light and provide color vision

---

a chemical, secreted into a synaptic cleft by a neuron, that affects nervous system signaling by binding with receptors on an adjacent neuron or effector cell

---

an endocrine gland that receives its hormones directly from the hypothalamus, then stores these hormones and later releases them

---

the sense of smell

---

automatic nervous system response that helps an organism avoid danger or preserve a stable physical state; the knee-jerk response is a well-known reflex

---

the division of the autonomic nervous system that generally has relaxing effects on the body

retina

sympathetic division

rods

synapse

sensory neuron

synaptic cleft

somatic nervous system

target cells

steroid hormones

thalamus

the division of the autonomic nervous system that generally has stimulatory effects on the body

a layer of tissue at the back of the human eye whose cells convert light signals into neural signals

an area in the nervous system consisting of a sending neuron, a receiving neuron or effector cell (i.e., a muscle cell), and the gap between them, called the synaptic cleft

photoreceptors that function in low-light situations but that do not provide color vision

tiny gap that exists between a neuron sending a nervous system signal and a neuron (or effector cell) that is receiving this signal

a neuron that senses conditions both inside and outside the body and conveys that information to neurons in the central nervous system

the set of cells in the body that will be affected by a given hormone; the target cells for each hormone bear receptors capable of binding with that hormone

that portion of the peripheral nervous system's efferent division that provides voluntary control over skeletal muscle

portion of the brain through which most sensory perceptions are channeled before being relayed to the cerebral cortex

hormones, such as testosterone and estrogen, that are constructed around the chemical framework of the cholesterol molecule; one of three principal classes of hormones, the other two being peptide hormones and amino acid-based hormones

SELF TEST

Once you have finished studying this chapter, close your books, grab a pencil, and spend the next 15 to 20 minutes completing this practice test.

Compare and Contrast

For each of the following paired terms, write a sentence of comparison ("Both") and a sentence of contrast ("However,").

nerve/neuron
steroid hormone/amino acid-based hormone
pituitary/hypothalamus
axon/dendrite
sympathetic/parasympathetic

Short Answer

1. Explain the difference in the communication methods of the nervous and endocrine systems.

2. Explain the relationship of rods and cones.

3. How is homeostasis maintained in the body?

4. Which type of gland secretes hormones, and what is unique about these glands?

5. Which type of hormone is derived from cholesterol? Can you see why a certain amount of cholesterol is necessary?

6. What effects do hormones have on cells?

7. Why do organisms rely more heavily on negative feedback than positive feedback?

8. As you turn your head, you smell a rose. You reach out to pick it up. Explain the roles of motor neurons, sensory neurons, and interneurons in this scenario.

9. Describe the ion flow associated with an action potential. How does an action potential travel down an axon?

10. What do neurotransmitters do? Do all neurotransmitters send the same message?

11. Why do you pull your hand back so quickly after being stuck by a pin? Describe the parts of the reflex arc that are required for that action to occur.

12. Hypoglycemia is a condition in which blood sugar drops to very low levels. What would seem to be the cause of hypoglycemia? Do you think this problem could be corrected by eating more sugar? Why might this not be such a good idea?

13. There are many different kinds and causes of deafness. Explain how the following would result in deafness: damage to hair cells from repeated loud sounds, brain injury, torn eardrum, and buildup of earwax in the auditory canal.

14. How are the sense of smell and sense of taste similar? What are the differences between the senses of taste and smell?

15. Why is it that a person can survive in a vegetative state after massive trauma to the cerebrum as long as the brain stem (thalamus, midbrain, pons, medulla oblongata) is intact?

16. The pituitary is often called the master gland. In what way is this true? In what way is this misleading?

Multiple Choice

Circle the letter that best answers the question.

1. Which of the following represents the correct direction of signal movement through a neuron?
   a. axon → cell body → dendrite
   b. dendrite → cell body → axon
   c. cell body → axon → dendrite
   d. axon → dendrite → cell body
   e. dendrite → axon → cell body

2. What common molecule forms the core of steroid hormones such as testosterone?
   a. collagen
   b. cellulose
   c. cholesterol
   d. calcium
   e. chitin

3. The brain and spinal cord are part of the:
   a. peripheral nervous system.
   b. autonomic nervous system.
   c. parasympathetic nervous system.
   d. sympathetic nervous system.
   e. central nervous system.

4. What are neurotransmitters used for?
   a. transmission of an action potential along an axon
   b. transmission of an action potential between a dendrite and an axon
   c. action potential transmission from an axon to a cell body
   d. action potential transmission between two neurons at a synapse
   e. to sense the environment

5. If you sustained an injury to your cerebellum, which of the following might you have difficulty with?
   a. breathing
   b. keeping balance
   c. studying
   d. sensing information from the external environment
   e. There would be no consequences of an injury to the cerebellum.

6. What controls the anterior pituitary?
   a. posterior pituitary
   b. antidiuretic hormone
   c. thalamus
   d. hypothalamus
   e. oxytocin

7. The brain knows that light has stimulated the rods and cones of the eye when:
   a. the rods and cones change color.
   b. only the cones increase the rate of signaling.
   c. only the rods increase the rate of signaling.
   d. the rods and cones stop signaling.
   e. the rods change color but the cones remain the same.

8. The complex that allows the transfer of an action potential from a sensory neuron to a motor neuron or interneuron is called a:
   a. neuroglia.
   b. neuropathy.
   c. synapse.
   d. complexion.
   e. cortex.

9. All of these are part of a reflex arc *except* the:
   a. sensor receptor.
   b. effector (muscle or gland).
   c. spinal cord.
   d. cerebral cortex.
   e. All of the above are required for a functional reflex.

10. Which of the following is the brain structure that allows you to perceive and answer this question?
    a. cerebellum
    b. pons
    c. hypothalamus
    d. cerebrum
    e. thalamus

11. Which of the following is a true statement about the endocrine system?
    a. Most hormones are made from carbohydrates.
    b. Most hormones act only where they are synthesized.
    c. Every hormone has a specific target that it affects.
    d. Neurotransmitters are DNA-based hormones.
    e. Endocrine glands are composed of mesothelial tissues.

12. Which of the following is the best example of negative feedback?
    a. Drinking salty seawater makes you thirsty, so you drink more.
    b. Uterine contractions during childbirth send signals to the brain that increase the rate and intensity of contractions.
    c. Eating one potato chip gives you cravings for more.

    d. Once a thermostat is set for a certain temperature, the furnace will shut off when it reaches that temperature.
    e. Heavy rains cause flooding.

13. Which of the following would be a homeostatic mechanism?
    a. receptors that make you feel more tired as you increase the number of hours you sleep each night
    b. receptors that cause an increase of blood flow to an area with an open wound
    c. receptors that respond to a drop of blood iron levels and send a message to the brain that causes you to crave spinach (an iron-rich food)
    d. receptors that respond to the color of spring flowers
    e. none of the above

Match the following terms with their descriptions by writing the appropriate letters in the blanks.

    a. autonomic nervous system
    b. hormone
    c. action potential
    d. synapse
    e. pituitary

14. _____ Message carried by an axon

15. _____ Control center of the endocrine system

16. _____ Site of message transfer in the nervous system

17. _____ Regulation of smooth muscle

18. _____ Message received via the bloodstream

19. Josh is color-blind, so he has a lot of trouble picking out clothes. Which sensory structures below are defective in a color-blind person?
    a. organs of Corti
    b. cones
    c. hair cells
    d. rods
    e. utricle and saccule

20. What is the difference between a neuron and a nerve?
    a. One is sensory in function, the other motor.
    b. Nerves are found only in the central nervous system.
    c. Neurons are made of white matter, nerves of gray matter.
    d. Neurons are found only in mammals.
    e. They each consist of a different numbers of cells.

21. Target cells:
    a. are found only in specific endocrine glands.
    b. are equipped with specific receptor molecules.
    c. are muscle cells.
    d. are found only in the brain.
    e. are found only when a hormone is secreted.

22. Blood glucose levels are regulated by:
  a. insulin and glucagon.
  b. insulin and cortisol.
  c. glucagon and cortisol.
  d. glucagon alone.
  e. insulin alone.

23. The part of the brain that controls basic responses necessary to maintain life processes (breathing, heartbeat) is the:
  a. medulla oblongata.
  b. cerebrum.
  c. cerebral cortex.
  d. cerebellum.
  e. spinal cord.

24. Which of the following is a junction between two neurons?
  a. synapse
  b. action potential
  c. myelin sheath
  d. gap junction
  e. sodium gate

25. The layer of the eye where the photoreceptors are located is the:
  a. lens.
  b. pupil.
  c. cornea.
  d. retina.
  e. iris.

WHAT'S IT ALL ABOUT?

Here's a question to help you pull together what you've learned so far using this text.

Question: The visual system has three component tasks, which simplify conceptually to "convert light signals into information we can use." Being able to use that information—to climb a stair, avoid a snake, pick an apple—is what matters to us. Occasionally, people who have lost their vision when very young regain it after surgery as adults. Despite being able to see, however, these people often find it harder to make sense of their world. Why?

What do I do now?
Remember the drill—decide what the question is asking you to do, collect your evidence from this chapter (and the others you've studied), and write!

CHAPTER 28   DEFENDING THE BODY: THE IMMUNE SYSTEM

Basic Chapter Concepts

* The immune system employs nonspecific and specific defenses to protect the body from foreign organisms and toxins. Nonspecific defenses block or attack invaders indiscriminately, whereas specific defenses target invaders and create a "memory" within the immune system to combat future attacks.

CHAPTER SUMMARY

28.1   Two Types of Immune Defense

* Nonspecific defenses of the immune system do not employ invader-specific mechanisms. Physical barriers to penetration by microorganisms, phagocytic cells, and antiviral compounds define most of the nonspecific "arsenal."
* Specific defenses employ T- and B-lymphocyte cells, and their products, to eliminate invading organisms. T and B cells recognize features of the invader in much the same way that receptors recognize their ligands. When an invader binds, a response is triggered, which can include the release of cytokines or the production of antibodies.

28.2   Nonspecific Defenses

* The nonspecific defenses can be divided into two groups: the first and second lines of defense. The first line includes the passive barriers of the skin and mucous membranes. The second line includes the phagocytic cells and the immune response.

28.3   Specific Defenses

* Immunity can be acquired passively or actively. Passive immunity comes from microorganism compounds that are made outside the body and introduced into the body for the short-term goal of eliminating the invader (antigen). Active immunity is long-lasting protection gained after a previous exposure to the invader, because of either previous illness or vaccination.
* Actively acquired immunity consists of the action of cell-mediated and antibody-mediated processes. Cell-mediated processes involve activated T cells that produce antigen-fighting chemicals or become phagocytotic.

28.4   Antibody-Mediated Immunity

* Antibody-mediated immunity relies on B cells to produce specific antibodies that bind to the antigen and drive its removal from the body. Both immune processes can build a memory, so that subsequent exposure to the antigen reactivates the immune response.

28.5   Cell-Mediated Immunity

* Cell-mediated immunity involves infected cells and the T cells that specialize in those infections. Helper T, regulatory T, and cytotoxic T cells all play roles in this level of defense.
* Activation of helper T cells by an APC cell causes the release of Interleukin-2. This in turn stimulates cloning of cytotoxic Ts. Cytotoxic Ts destroy infected cells by creating holes in the cell membrane and, additionally, stimulating apoptosis. As this process is going on, memory cell versions of the cytotoxic T and helper T cells form, ensuring immunity to a second exposure of the pathogen. Regulatory T cells are activated as well. These in turn shut down or limit the immune response.

28.6 AIDS: Attacking the Defenders

- In AIDS (acquired immunodeficiency disease syndrome) the human immune system has been attacked by the invading virus, thus eliminating the system that could defeat the virus.
- Development of a vaccine for HIV is very difficult because it could potentially interfere with the function of the helper T cells which are the targets of the virus. In addition, the virus is capable of rapid mutation making it a hard target to attack by conventional methods of vaccine development.

28.7 The Immune System Can Cause Trouble

- Occasionally the body mistakes its own tissues for foreign antigens and elicits an immune response against itself. Allergic reactions are one form of autoimmune response; rheumatoid arthritis, lupus, and diabetes are others. Autoimmune responses can be deadly.

WORD ROOTS

anti- = against (e.g., *anti*bodies are proteins that defend the body against pathogens)

patho- = disease (e.g., *patho*logy is the study of disease)

phago- = to eat (e.g., *phago*cytes are cells that engulf (eat) and destroy materials or cells)

KEY TERMS

actively acquired immunity _____

_____

allergen _____

_____

allergy _____

_____

antibody _____

_____

antibody-mediated immunity _____

_____

antigen _____

_____

antigen-presenting cell (APC) _____

_____

autoimmune disorder _____

_____

B-lymphocyte cell (B cell) _____

_____

cell-mediated immunity _____

_____

cytotoxic T cell (killer T cell) _____

_____

helper T cell _____

_____

histamine _____

_____

HIV _____

_____

immunity _____

_____

memory B cell _____

_____

nonspecific defenses _____

_____

passively acquired immunity _____

_____

phagocyte _____

_____

plasma cell _____

_____

regulatory T cell _____

_____

specific defenses _____

_____

T-lymphocyte cell (T cell) _____

_____

FLASH CARDS

To use the flash cards, tear the page from the book and cut along the dotted lines. The key term appears on one side of the flash card, and its definition appears on the opposite side.

actively acquired immunity	autoimmune disorder
allergen	B-lymphocyte cell (B cell)
allergy	cell-mediated immunity
antibody	cytotoxic T cell (killer T cell)
antibody-mediated immunity	helper T cell
antigen	histamine
antigen-presenting cell (APC)	HIV

an attack by the immune system on the body's own tissues

immunity developed as a result of accidental or deliberate exposure to an antigen; accidental exposure: coming into contact with someone who has a transmissible disease; deliberate exposure: being vaccinated

the central cells of antibody-mediated immune system function; B cells produce antibodies, called antigen receptors, that bind with specific antigens while remaining embedded in the B cell; conversely, these antibodies may be exported from B cells as free-standing entities to fight antigens

a foreign substance that triggers an allergic reaction; these substances are usually derived from living things, including pollen, dust mites, foods, and fur

an immune system capability that works through the production of cells that destroy infected cells in the body

an overreaction by the immune system to an antigen, resulting in the release of histamine

type of T-lymphocyte cell that binds to and kills the body's own cells when they have become infected

a protein of the immune system that is found on the surface of B cells, or that is exported by them, that is able to bind to a specific antigen, thus playing a role in eliminating it from the body

type of T-lymphocyte cell that stimulates both T-cell and B-cell immunity; referred to in AIDS therapy as a CD-4 cell

an immune system capability that works through the production of proteins called antibodies

a compound in the immune system's inflammatory response that brings about blood vessel dilation and increased blood vessel permeability

any foreign substance that elicits a response by the immune system; certain proteins on the surface of an invading bacterial cell, for instance, act as antigens that trigger an immune response

human immunodeficiency virus, the cause of the disease AIDS

any immune system cell that presents, on its surface, fragments of an antigen that it has ingested; dendritic cells, macrophages, and B cells are the immune system's three classes of antigen-presenting cells

immunity

plasma cell

memory B cell

regulatory T cell

nonspecific defenses

specific defenses

passively acquired immunity

T-lymphocyte cell (T cell)

phagocyte

one of two types of cells the B-lymphocyte cell develops into (the other type being a memory B cell); plasma cells are specialized to produce the free-standing antibodies that fight invaders

a state of long-lasting protection that the immune system develops against specific microorganisms

a type of immune system cell that acts to limit the body's immune system response, thus protecting the body's own tissues from attack

one of two types of cells that the B-lymphocyte cell, or B cell, develops into (the other type being a plasma cell); memory B cells remain in the system long after a first infection by a microorganism has ended and serve to produce more plasma B cells quickly should the microorganism ever invade again

immune system defenses that provide protection against particular invaders

immune system defenses that do not discriminate between one invader and the next

a class of lymphocytes that plays a central role in cell-mediated immunity; T cells come in several varieties that play specific roles in recognizing and killing infected cells in the body

immunity gained by the administration of antibodies produced by another individual, such as by the administration of a gamma globulin shot

an immune system cell capable of ingesting another cell, parts of cells, or other materials; phagocytes ingest both invading microorganisms and tissue fragments or the body's own cells when they have become damaged

SELF TEST

After you have finished studying this chapter, close your books, grab a pencil, and spend the next 15 to 20 minutes completing this practice test.

Compare and Contrast

For each of the following paired terms, write a sentence of comparison ("Both") and a sentence of contrast ("However").

specific defense/nonspecific defense
B cell/T cell
histamine/cytokine
dendritic cells/macrophages
plasma cells/B cells

Short Answer

1. What are autoimmune disorders?
2. What is a vaccine?
3. What is the inflammatory response?
4. Explain the difference between polio and AIDS.
5. Why are antigen-presenting cells (APCs) so important to the immune response?
6. List the three types of T cells, and explain the function of each.
7. How do antibodies help the body to fight infection?

Multiple Choice

Circle the letter that best answers the question.

1. Which of the following is *not* a molecule used by the nonspecific defenses?
   a. lysozyme
   b. allergen
   c. complement
   d. histamine
   e. None of the above is used by the nonspecific defenses.

2. How does a natural killer cell recognize cancer cells?
   a. by their location
   b. by their accelerated growth rate
   c. by their altered cell surface proteins
   d. by their mutated DNA
   e. none of the above

3. What is the major benefit of the specific defense system?
   a. Specific defense systems act as barriers to foreign invaders.
   b. Specific defenses provide a quicker response than nonspecific defenses.
   c. Specific responses are generated no matter what the situation is.
   d. Specific defenses can produce immunity.
   e. All immune responses are specific.

4. Which of the following cells engulf and digest other cells from your body?
   a. helper T cells
   b. suppressor T cells
   c. B cells
   d. macrophages
   e. natural killer cells

5. What do rheumatoid arthritis and multiple sclerosis have in common?
   a. They are both diseases of the joints.
   b. They are both diseases of the nervous system.
   c. They are both forms of small-cell cancer.
   d. They are both the result of malfunctions of the immune system.
   e. They are both caused by bacterial infections.

6. What system is attacked by the human immunodeficiency virus?
   a. nervous
   b. reproductive
   c. nonspecific immune
   d. specific immune
   e. the blood-cell-generating bone marrow

7. Serafina is fighting off an infection. She has managed to drag herself over to visit you despite the fever and other symptoms she is suffering. She vows to have a doctor cut out her swollen, painful lymph glands and thereby remove the source of her illness. You explain that it would be a bad idea, because:
   a. the glands are filled with immune-system cells fighting the infection.
   b. those little pathogens get really mad when you destroy their dens.
   c. lymph glands are found only in the central portion of the cerebral cortex in the brain.
   d. removing the glands will cause the infection to spread.
   e. lymph glands actually have little to do with the course of infection.

8. Activation of which of the following is required for the production of antibodies?
   a. B-lymphocyte cells
   b. helper T cells
   c. plasma cells
   d. antigens
   e. All of these play a role in generating antibodies.

9. Which of the following is *not* a symptom of an inflammation?
   a. redness
   b. chills
   c. pain
   d. swelling
   e. heat

10. Which of the following is *not* a type of T cell?
    a. killer
    b. cytotoxic
    c. helper
    d. inhibitory
    e. regulatory

11. Passive immunity:
    a. can be passed from mother to child.
    b. can be generated only for bacterial infections.
    c. can be administered as an injection.
    d. can be found only in cows (vacca).
    e. both a and c

12. What actually stimulates helper T cells to begin the specific immune response?
    a. allergens
    b. cytokines
    c. antibodies
    d. histamine
    e. lymph

13. Which of the following are activated along with natural killer cells and macrophages during the nonspecific immune response?
    a. complement proteins
    b. allergens
    c. antigens
    d. dendritic cells
    e. B cells

14. Which of the following causes the release of histamine?
    a. complement proteins
    b. allergens
    c. antigens
    d. dendritic cells
    e. B cells

15. Which of the following present antigens to other cells?
    a. complement proteins
    b. allergens
    c. antigens
    d. dendritic cells
    e. B cells

16. Which of the following generate the formation of antibodies?
    a. complement proteins
    b. allergens
    c. antigens

   d. dendritic cells
   e. B cells

17. Which of the following stimulate both B cells and T cells?
    a. complement proteins
    b. allergen
    c. helper T cells
    d. dendritic cells
    e. B cells

18. Which of the following give rise to the formation of plasma cells?
    a. allergens
    b. helper T cell
    c. antigens
    d. dendritic cells
    e. B cells

19. Which of the following are infection targets for HIV?
    a. macrophages
    b. dendritic cells
    c. helper T cells
    d. cytotoxic T cells
    e. B cells

20. HIV is a particularly damaging pathogen for several reasons. Which of the following is *not* one of them?
    a. It attacks the immune system itself.
    b. Its invasion process leads to a harmful infection very quickly.
    c. It survives for long periods of time outside the body.
    d. Its genetic material mutates rapidly.
    e. It has no cure.

21. Place the following in the correct order of activation:
    i. B and T cells
    ii. antigen-presenting cells
    iii. plasma cells
    iv. helper T cells
    a. iii, ii, i, iv
    b. ii, i, iv, iii
    c. iv, ii, i, iii
    d. iii, i, ii, iv
    e. ii, iv, i, iii

22. To produce antibodies, an immune cell must increase its production of:
    a. fatty acids.
    b. proteins.
    c. carbohydrates.
    d. lipid backbones.
    e. nucleic acids.

23. Which of the following is *not* true about the lymphatic system?
    a. It collects interstitial fluid.
    b. It includes lymph nodes, which harbor lymphocytes.

c. It is completely separate from the vascular system.

d. It is found throughout the body.

e. It includes the spleen.

24. Dendritic cells are most similar in action to:

   a. macrophages.

   b. helper T cells.

   c. B cells.

d. cytotoxic T cells.

e. natural killer cells.

25. Memory cells are:

   a. required for immunity.

   b. derived from macrophages.

   c. the source of histamine.

   d. antigen-presenting cells.

   e. first responders in active immunity.

WHAT'S IT ALL ABOUT?

Here's a question to help you pull together what you've learned so far using this text.

Question: In recent times, we have seen an increase in the number of people suffering from asthma and allergies, especially young children. Some people have suggested that this has happened because the immune system was not adequately stimulated early in life; that is, toddlers aren't coming into contact with enough "dirt" because of our air-conditioned homes, antibacterial cleaners, and limited outside play. Given what you know about specific immunity, is there any support for this hypothesis?

What do I do now?

Remember the drill—decide what the question is asking you to do, collect your evidence from this chapter (and the others you've studied), and write!

CHAPTER 29   TRANSPORT AND EXCHANGE 1: BLOOD AND BREATH

Basic Chapter Concepts

- The cardiovascular system transports gases, nutrients, and waste products throughout the body using muscular contractions of the heart.
- The respiratory system functions primarily to supply oxygen to the body's tissues and remove carbon dioxide, but it also provides immune surveillance, regulates blood pressure and pH, and provides structures for sound generation.

CHAPTER SUMMARY

29.1   The Cardiovascular System
- The cardiovascular system is the fluid transport system of the body, moving nutrients, vitamins, waste products, hormones, and cells of the immune system throughout the body. These functions are carried out by a system of vessels through which blood flows, and the propulsion for the entire system comes from the heart.

29.2   The Composition of Blood
- Blood consists of cells suspended in a protein-rich plasma. The red blood cells transport oxygen and carbon dioxide, and the white blood cells perform immune functions. The plasma transports hormones and fatty acids bound to proteins and carries nutrients and wastes dissolved in solution.

29.3   Blood Vessels
- Blood vessels have an inner layer of epithelial cells, a middle layer of smooth muscle, and an outer layer of connective tissue. Vessels that carry blood away from the heart are arteries and arterioles; veins return blood to the heart. Valves maintain unidirectional flow in the veins.

29.4   The Heart and Blood Circulation
- The heart sits at the center of two circulation systems—pulmonary and systemic. In the pulmonary circulation, blood is pumped into the lungs by the right ventricle, and oxygenated blood from the lungs is received into the left atrium. In the systemic system, blood from the left atrium flows into the left ventricle and is pumped out to the body. Blood returning from the body enters the right atrium through the venae cavae.

29.5   What Is a Heart Attack?
- Like any tissue, the heart is surrounded by coronary arteries that supply this muscle with oxygen and nutrients. Blockages within those arteries starve the heart tissue of essential nutrients and cause tissue death; this is a myocardial infarction—a heart attack.

29.6   Distributing the Goods: The Capillary Beds
- Nutrients and gases are delivered to the tissues in the capillary beds. Pressure resulting from the heart's contractions pushes gases and small molecules through the capillary walls and into the interstitial fluids. Osmotic pressure works to move water and small molecules (urea) back into the veins.

29.7   The Respiratory System
- Oxygen is captured and carbon dioxide is removed in the alveoli of the lungs. The lungs are composed of highly branched passages—the bronchi and bronchioles—to maximize the surface area available for gas exchange with the atmosphere. Structures that move gases from inside the body to outside include the nose, pharynx, and trachea.

29.8   Steps in Respiration
- Air moves into and out of the lungs because of changes in air pressure during the breathing (respiratory) cycle. As the diaphragm expands (moves down), the chest volume increases; as a result, the internal air

pressure decreases, drawing air into the body. During exhalation, the diaphragm relaxes, decreasing chest volume, increasing air pressure, and emptying the lungs.
- Respiration involves gas exchange—exchanging gases in the blood with the outside world through the lungs and exchanging gases between the blood and the interstitial fluid of the tissues. Both processes are driven by diffusion down concentration gradients.
- Oxygen and carbon dioxide are transported to and from tissues by the red blood cells. Oxygen travels bound to hemoglobin, whereas most of the carbon dioxide is transported as bicarbonate within the red blood cell.

WORD ROOTS

cardio- = heart (e.g., the *cardio*vascular system is a system of the body consisting of the heart, all the blood vessels in the body, and the blood that flows through these vessels)

heme = iron-containing compound (e.g., *hemo*globin is a molecule consisting of four protein subunits, each containing an iron atom bound to a heme group)

lipo- = fat, or fatty tissue (e.g., high-density *lipo*protein is a type of protein that transports fat, or lipid, molecules)

KEY TERMS

alveoli _____

aorta _____

artery _____

bronchiole _____

capillary _____

cardiovascular system _____

coronary artery _____

formed elements _____

heart attack _____

hemoglobin _____

high-density lipoprotein (HDL) _____

low-density lipoprotein (LDL) _____

plasma _____

platelet _____

pulmonary circulation _____

red blood cell (RBC) _____

respiration _____

systemic circulation _____

vein _____

ventilation _____

white blood cell (WBC) _____

FLASH CARDS

To use the flash cards, tear the page from the book and cut along the dotted lines. The key term appears on one side of the flash card, and its definition appears on the opposite side.

alveoli	coronary artery
aorta	formed elements
artery	heart attack
bronchiole	hemoglobin
capillary	high-density lipoprotein (HDL)
cardiovascular system	low-density lipoprotein (LDL)

an artery that delivers oxygenated blood to the muscles of the heart; blockage of coronary arteries causes heart attack

tiny, hollow air-exchange sacs that exist in clusters at the end of each of the air-conducting passageways in the lungs, the bronchioles

cells and cell fragments that form the nonfluid portion of blood; formed elements include red blood cells, white blood cells, and platelets; contrast with plasma, the fluid portion of blood, consisting mostly of water

the enormous artery extending from the heart that receives all the blood pumped by the heart's left ventricle; branches stemming from the aorta supply oxygenated blood to all the tissues in the body

a complete blockage of one of the heart's coronary arteries, resulting in the death of groups of heart cells from lack of a blood supply

a blood vessel that carries blood away from the heart

the iron-containing protein in red blood cells that binds to both oxygen and carbon dioxide, thus assisting in their transportation

tiny air-conducting passageway in the lungs that has at its end several alveoli, the hollow air-exchange sacs of the lungs

a type of protein that transports fat, or lipid, molecules (usually cholesterol) from various tissues in the body to the liver; sometimes known as the "good cholesterol," HDLs help remove, and possibly neutralize, the LDLs or low-density lipoproteins that can harmfully reside in coronary arteries

the smallest type of blood vessel, connecting the arteries and veins in the body's tissues; gases, nutrients, and wastes are exchanged between the blood and the body's tissues through the thin walls of capillaries

a type of protein that transports fat, or lipid, molecules (usually cholesterol) from the liver and small intestine to various tissues throughout the body; sometimes known as the "bad cholesterol," LDLs can initiate heart disease by coming to reside within coronary arteries

a fluid transport system of the body, consisting of the heart, all the blood vessels in the body, the blood that flows through these vessels, and the bone marrow tissue in which red blood cells are formed

plasma

systemic circulation

platelet

vein

pulmonary circulation

ventilation

red blood cell (RBC)

white blood cell (WBC)

respiration

one of the body's two general networks of blood vessels that serve to transport blood between the heart and the rest of the body following oxygenation of this blood in the body's pulmonary circulation

the fluid portion of blood, consisting mostly of water but also containing proteins and other molecules; contrast with formed elements—the cells and cell fragments that form the nonfluid portion of blood

a blood vessel that carries blood toward the heart

one of the three varieties of formed elements within blood, platelets are small fragments of cells that facilitate blood clotting by releasing clotting enzymes and clumping together at the site of an injury

the physical movement of air into and out of the lungs, brought about by contractions and relaxations of muscles in the chest

the system that circulates blood between the heart and the lungs; this system brings oxygen into and takes carbon dioxide away from the body

the central cells of the immune system; types of white blood cells include T cells, B cells, neutrophils, eosinophils, and macrophages, each playing a specific role in immune responses

the blood cells, also known as erythrocytes, that transport oxygen to and carry carbon dioxide from every part of the body

the movement of oxygen from outside the cell to inside the cell, and the movement of carbon dioxide from inside the cell to outside the cell

SELF TEST

Once you have finished studying this chapter, close your books, grab a pencil, and spend the next 15 to 20 minutes completing this practice test.

Compare and Contrast

For each of the following paired terms, write a sentence of comparison ("Both") and a sentence of contrast ("However,").

red blood cells/white blood cells
artery/vein
high-density lipoprotein/low-density lipoprotein
respiration/ventilation
alveoli/bronchiole

Short Answer

1. How does the cross section of a capillary compare to the cross section of an artery? How does it compare to the cross section of a vein?

2. Why does the human heart contain four chambers? List these chambers and describe the function of each. What is the order of blood flow through the heart?

3. Which circulatory system picks up oxygen from the lungs, and which circulatory system drops off oxygen at the tissues of the body?

4. What vessels are blocked in a heart attack?

5. Where does exchange of gases between blood and cells of the body take place? What other system assists in returning blood back toward the heart?

6. Why is osmosis important in the exchange of nutrients between blood and the cells of the body?

7. How do humans breathe, and what muscles are involved?

8. When spread out, the surface of alveoli in the human lung is said to approximate that of a tennis court. Why do you think this surface area needs to be so large?

Multiple Choice

Circle the letter that best answers the question.

1. Which chamber of the heart receives blood from the lungs?
   a. right atrium
   b. right ventricle
   c. left ventricle
   d. left atrium
   e. pulmonary vein

2. Which solid element of blood is involved in clotting?
   a. red blood cells
   b. white blood cells
   c. platelets
   d. erythrocytes
   e. leukocytes

3. White blood cells:
   a. assist in blood clotting.
   b. transport oxygen.
   c. remove carbon dioxide from the blood.
   d. are active in defending the body against invaders.
   e. are red blood cells that lack melanin.

4. The left ventricle is larger and stronger than the right ventricle because:
   a. the left is the first chamber that the blood enters as it returns to the body.
   b. the left is the chamber that must pump blood to the lungs.
   c. the left is the chamber that must pump blood to the body.
   d. all the structures on the left side of the body are larger than those on the right.
   e. b and d

5. Which of the following is the site of exchange for oxygen and carbon dioxide?
   a. pharynx
   b. trachea
   c. bronchioles
   d. alveoli
   e. esophagus

6. A heart attack is caused by:
   a. a backup of bile in the liver.
   b. a spasm during peristalsis in the duodenum.
   c. lack of blood flow to cardiac muscle cells.
   d. an autoimmune disease.
   e. none of the above

7. Ninety-two percent of plasma is made up of:
   a. red blood cells.
   b. white blood cells.
   c. platelets.
   d. fibrinogen.
   e. water.

8. The illegal practice of blood doping is dangerous because:
   a. it causes an abnormal drop in red blood cells.
   b. instead of increasing oxygen delivery to cells, it actually decreases oxygen delivery to cells.
   c. it causes an abnormal increase in red blood cells, which can literally turn blood into sludge.

d. erythropoietin has not been approved for use by the FDA.

e. it increases the amount of circulating white blood cells, which can lead to infections.

9. Inhaled air passes last through which of the following?
   a. bronchiole
   b. larynx
   c. pharynx
   d. trachea
   e. bronchus

10. Heart valves function to:
    a. keep blood moving forward through the heart.
    b. mix blood thoroughly as it passes through the heart.
    c. control the amount of blood pumped by the heart.
    d. slow blood down as it passes through the heart.
    e. propel blood as it passes through the heart.

11. Blood moves most slowly in:
    a. capillaries.
    b. the aorta.
    c. veins.
    d. arterioles.
    e. venules.

12. Which of the following characteristics best describes an artery?
    a. carries blood away from the heart
    b. carries oxygenated blood
    c. contains a valve
    d. has thin walls
    e. carries blood away from capillaries

13. Most of the oxygen in the blood is transported by:
    a. plasma.
    b. serum.
    c. platelets.
    d. hemoglobin.
    e. leukocytes.

14. The atria of the heart differ from the ventricles in that they:
    a. are larger.
    b. have thicker walls with more muscles.
    c. receive blood from veins.
    d. have a higher blood pressure.
    e. empty through valves.

15. The diastolic pressure for a normal young adult is:
    a. 60 mm Hg.
    b. 80 mm Hg.
    c. 100 mm Hg.
    d. 120 mm Hg.
    e. 140 mm Hg.

16. The movement of both oxygen and carbon dioxide in the body is accomplished by:
    a. simple diffusion.
    b. osmosis.
    c. cell movement.
    d. ciliated cells.
    e. the diaphragm.

17. Actual exchange of gases in the lungs occurs in the:
    a. bronchi.
    b. bronchioles.
    c. alveoli.
    d. trachea.
    e. diaphragm.

18. Other than water, plasma also contains:
    a. red blood cells.
    b. albumins, globulins, and fibrinogen.
    c. white blood cells.
    d. albumins and fatty acids.
    e. platelets and fibrinogen.

19. The coronary vessels:
    a. supply and drain the heart muscle.
    b. bypass the ventricles.
    c. send blood directly to the lungs.
    d. are not really necessary, because the heart gets its blood from inside the heart.
    e. lead directly from the atria to the lungs.

20. Cholesterol is carried by:
    a. albumin.
    b. globulins.
    c. lipoproteins.
    d. triglycerides.
    e. fibrinogen.

21. The "frontier" through which exchange of substances between blood and surrounding cells occurs is through the walls of the:
    a. arterioles.
    b. capillaries.
    c. arteries.
    d. veins.
    e. heart muscle.

22. Which of the following is *not* a function of blood?
    a. transport oxygen and carbon dioxide
    b. act as a cooling and heating fluid for the body
    c. produce red blood cells
    d. transport nutrients
    e. transport components of the immune system

23. The familiar heart sounds that are heard with a stethoscope are produced by the:
    a. contraction of cardiac muscle.
    b. nerve impulses of the heart.
    c. blood rushing against closing heart valves.

d. leakage of some blood back into the heart with each beat.

e. blood flowing through the coronary artery.

24. A structure that greatly increases the surface area for respiratory exchange in the human lung is the:
   a. bronchiole.
   b. alveolus.
   c. trachea.
   d. pharynx.
   e. nasal cavity.

25. A molecule of carbon dioxide that is released into your blood from your right big toe is exhaled out of your nose. It must pass through all of the following structures *except:*
   a. the right atrium.
   b. a pulmonary vein.
   c. an alveolus.
   d. the trachea.
   e. the bronchi.

LABELING

Label the figure with the following terms: right ventricle, left ventricle, pulmonary arteries, pulmonary veins, aorta, left atrium, right atrium.

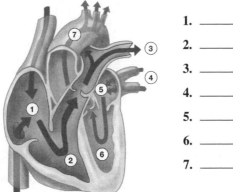

1. _____
2. _____
3. _____
4. _____
5. _____
6. _____
7. _____

WHAT'S IT ALL ABOUT?

Here's a question to help you pull together what you've learned so far using this text.

Question: People with "poor circulation" suffer a variety of problems. Their fingers and toes may get cold easily. They are easily tired out by climbing stairs. They may suffer more frequent infections. They may seem lethargic. How are these ills connected to circulation?

What do I do now?
Remember the drill—decide what the question is asking you to do, collect your evidence from this chapter (and the others you've studied), and write!

CHAPTER 30   TRANSPORT AND EXCHANGE 2: DIGESTION, NUTRITION, AND ELIMINATION

Basic Chapter Concepts

- The digestive system extracts usable material from ingested food and disposes of the waste.
- The urinary system disposes of metabolic wastes and maintains fluid balance in the body.

CHAPTER SUMMARY

30.1   The Digestive System
- Components of the digestive system make small molecules out of big molecules, so that nutrients can diffuse into the circulation and be delivered to the tissues for use in synthetic reactions.

30.2   Structure of the Digestive System
- The order of major structures in the digestive tract is as follows: mouth, pharynx, esophagus, stomach, small intestine, and large intestine.
- The four layers of tissue found in the majority of these structures are as follows: the serosa (outermost), the muscularis (longitudinal and circular layers of muscle), and the submucosa (nerve, vascular, and mucus-producing cells).
- The muscularis externa is responsible for the movement of material through the digestive tract known as peristalsis.
- The liver, gallbladder, and pancreas are accessory organs that aid in digestion.

30.3   Steps in Digestion
- Different types of foods are degraded in different locations within the digestive tract. Carbohydrate digestion begins in the mouth and is completed in the small intestine; minimal degradation takes place in the stomach. Protein degradation is accomplished primarily by the gastric juice of the stomach. Fats are not digested until they enter the small intestine, where they are solubilized by bile salts and at this point serve as suitable substrates for the pancreatic lipase.
- Digestion consists of six steps. The first two, ingestion and mechanical degradation, occur within the oral cavity, pharynx, and esophagus. Food enters the mouth and is crushed and torn by the teeth. Saliva lubricates, and thus facilitates, the mechanical breakdown of the food. Enzymes within saliva begin the digestive process. The slurry of food now passes through the pharynx to the esophagus to make the trip to the stomach. The esophagus pushes the food down by waves of strong muscle contractions.
- The third and fourth steps of digestion occur within the stomach, where the macerated food is mixed with acid and enzymes, forming chyme. Gastric glands secrete a mixture of hydrochloric acid and the enzyme pepsin to facilitate breakdown of the proteins in food particles into amino acids. Muscular contractions of the stomach mix the chyme and gastric juice, accelerating the degradation.
- As the chyme enters the small intestine, additional enzymes secreted by the pancreas and bile secreted by the liver and gallbladder enter the digestive tract to complete the process of breakdown. The fifth step of digestion, absorption of small molecules into the capillaries and lymphatics, occurs almost exclusively in the small intestine.
- The final step of digestion, removal of waste products, occurs through the action of the large intestine. Water is absorbed by the large intestine, compacting the waste products. The rectum, a muscular chamber at the end of the large intestine, expels the fecal material.

30.4   Human Nutrition
- Nutrients are substances that are acquired through the diet to provide energy, regulatory structures, or essential building blocks for the body.
- Nutrients are divided into six classes; water, minerals, vitamins, proteins, lipids, and carbohydrates.

30.5   Water, Minerals, and Vitamins
- Water is required within the body to maintain fluid volume and cellular concentrations. The body is made up of about 66 percent water, and water is easily lost through urination, perspiration, and breathing; thus, water intake is essential to life.
- Minerals are used to build structures and/or enhance metabolic processes. Mineral deficiencies can make it harder for the body to build and maintain structures such as bone.
- Vitamins help enzyme function, and even though they may be required only in small quantities, they are also essential to life.

30.6   Calories and the Energy-Yielding Nutrients
- The Calorie (or kcal) is a measure of heat. One kcal is the amount of energy required to raise 1 liter (1,000 g) of water 1 degree Celsius.
- Proteins and carbohydrates produce about 4 kcal of energy per gram, and lipids produce about 9 kcal per gram.

30.7   Proteins
- There are 20 types of amino acids that are used to build proteins. Of those, 11 can be manufactured within the human body; the rest are essential amino acids; that is, they must be acquired through the diet.
- Animal sources provide all the essential amino acids. To provide the human body with the essential amino acids in the required proportions, plant sources must either be mixed (e.g., combination of grains and legumes) or be of certain specific types (e.g., soy).

30.8   Carbohydrates
- Carbohydrates can be divided into simple sugars (one to two units long) and complex carbohydrates (long chains). Complex carbohydrates include starches and cellulose, which comprises the indigestible "fiber" in the diet.
- Carbohydrates are used primarily as energy. One type in particular, glucose, is the preferred energy source for brain cells.

30.9   Lipids
- Lipids are made up of fatty acids: a hydrocarbon chain with an "acid" carboxyl group at its head. In the body, lipids are used for energy storage and cell membrane structure (phospholipids).
- Lipids can be subdivided as fats and oils. Although primarily considered as energy-rich molecules, certain types of lipids are essential to a healthy diet. Unsaturated omega-3 fats are important in preventing cardiovascular disease.
- Trans fats may actually promote cardiovascular disease.

30.10   Elements of a Healthy Diet
- Although there are a variety of ways to describe a healthy diet, food pyramids are particularly effective because they depict both content and quantity.

30.11   The Urinary System in Overview
- The urinary system—two kidneys, a bladder, and connecting structures—regulates fluid homeostasis. In addition to eliminating waste products, the kidneys maintain a constant blood volume and electrolyte concentration in the body.

30.12   Structure of the Urinary System
- The urinary system is made up of two kidneys and their attached ureters, which deliver urine to the urinary bladder for storage and eventual excretion.
- Within each kidney, the functional unit is the nephron. Each nephron is made up of the kidney tubule (functionally and structurally subdivided into sections) and the associated blood supply.

30.13   How the Kidneys Function
- Kidneys remove waste products from the blood in the nephrons. Blood enters Bowman's capsule under pressure, forcing water and small molecules, including the waste product urea, out of the blood and into

the proximal tubule. As this filtrate travels through the tubule, amino acids and sugars are actively pumped out of the tubule back into the blood, and the ions and water follow to maintain osmotic and electrical balance. Water continues to be removed from the filtrate in the distal part of the tubule and the loop of Henle, as ions and toxins move in. By the time the filtrate arrives at the collecting duct, it has become a concentrated solution of urea called urine.

30.14 Urine Storage and Excretion

- Urine travels to the urinary bladder through the ureters. The bladder holds up to 8 milliliters (mL), although the brain begins to receive signals from the stretch receptors signaling a need to urinate at 2 mL. Urine passes out of the bladder through the urethra, which has bands of smooth muscle (sphincters) at the top and bottom to control the flow of urine out of the bladder. Only the lower sphincter is under voluntary control.

WORD ROOTS

sacch- = referring to sugar or carbohydrates (e.g., a mono*sacch*aride is a single unit or simple sugar)

glyco- = referring to carbohydrates (e.g., *glyco*gen is a complex carbohydrate)

renal = referring to the kidney (e.g., the *renal* arteries carry blood to the kidney)

KEY TERMS

antidiuretic hormone (ADH) _____

bile _____

calorie (nutritional) _____

chyme _____

digestive tract _____

essential amino acids _____

fat _____

fibers _____

gallbladder _____

glomerulus _____

glycemic load _____

kidneys _____

large intestine _____

liver _____

mineral _____

monounsaturated fatty acid _____

nephron _____

nonessential amino acids _____

nutrient _____

nutrition _____

oil _____

pancreas _____

peristalsis _____

pharynx _____

phytochemical _____

polyunsaturated fatty acid _____

saturated fatty acid _____

_____

simple sugars _____

_____

small intestine _____

_____

starches _____

_____

stomach _____

_____

ureters _____

_____

urethra _____

_____

urinary bladder _____

_____

vitamin _____

_____

FLASH CARDS

To use the flash cards, tear the page from the book and cut along the dashed lines. The key term appears on one side of the flash card, and its definition appears on the opposite side.

antidiuretic hormone (ADH)	**fibers**
bile	**gallbladder**
calorie (nutritional)	**glomerulus**
chyme	**glycemic load**
digestive tract	**kidneys**
essential amino acids	**large intestine**
fat	**liver**

in skeletal muscle, a single elongated muscle cell containing hundreds of long, thin myofibrils that run the length of the cell; in nutrition, one of the three principal classes of dietary carbohydrate, defined as a complex carbohydrate that is indigestible; the other classes of dietary carbohydrate are simple sugars and starches

substance that helps control how much water is either sent to the bladder (in urine) by the kidneys or retained in circulation; in a release controlled by the brain's hypothalamus, ADH increases the permeability of both the distal nephron tubule and the nephron collecting duct to water, thus conserving it

organ of the body that stores and concentrates the digestive material bile, which is produced by the liver; bile facilitates the breakdown of fats by digestive enzymes

substance produced by the liver that facilitates the digestion of fats; bile can be released either directly by the liver or by the gallbladder, which stores and concentrates this substance

knotted network of capillaries in each of the kidneys' nephrons that receives blood and lets some smaller blood-borne materials pass out of it (and into the surrounding Bowman's capsule) while retaining larger materials

the amount of energy necessary to raise the temperature of 1,000 grams of water by 1°C; sometimes written as Calorie to distinguish it from a standard calorie, which is defined as the amount of energy needed to raise the temperature of 1 gram of water by 1°C

a measure of how blood glucose levels are affected by defined portions of given carbohydrates; lower glycemic loads are associated with improved human health

the soupy mixture of food and gastric juices that passes from the stomach to the small intestine

the filtering organs of the urinary system that produce urine while conserving useful blood-borne materials

a muscular passageway for food and food waste that runs through the human body from the mouth to the anus

that portion of the digestive tract that begins at the small intestine and ends at the anus; the large intestine serves mainly to compact and store material left over from the digestion of food, turning this material into the solid waste known as feces

in nutrition, one of nine amino acids that the body cannot make and that hence must be supplied by food

organ that is central to the body's metabolism of nutrients and that serves as a major storage site for blood

a dietary lipid that is solid at room temperature (e.g., butter, the fat in a piece of bacon)

mineral

pancreas

monounsaturated fatty acid

peristalsis

nephron

pharynx

nonessential amino acids

phytochemical

nutrient

polyunsaturated fatty acid

nutrition

saturated fatty acid

oil

simple sugars

in digestion, a gland that secretes, into the small intestine through ducts, digestive enzymes along with buffers that raise the pH of chyme; in nutrient metabolism, a gland that secretes, directly into the bloodstream, the hormones insulin and glucagon, which regulate blood levels of glucose

an element essential to the functioning of a living organism

waves of contraction carried out by two sets of muscles in the digestive tract that help digest material and push it through the tract

a fatty acid with one double bond between the carbon atoms of its hydrocarbon chain

in humans, the passageway at the back of the mouth that links the mouth with both the food-transporting esophagus and the air-transporting trachea; in some animals, the pharynx can be everted, or turned inside-out, and used to obtain nutrients

functional unit of the kidneys, composed of a nephron tubule, its associated blood vessels, and the interstitial fluid in which both are immersed

a nonnutritive substance found in plants that promotes health

in nutrition, one of the 11 amino acids that can be produced by the body and hence does not need to be supplied by food

a fatty acid with two or more double bonds between the carbon atoms of its hydrocarbon chain

a chemical element that is used by living things to sustain life

a fatty acid with no double bonds between the carbon atoms of its hydrocarbon chain

the study of the relationship between food and health

the smallest and simplest form of carbohydrates, which serve as energy-yielding molecules and as the building blocks or monomers of complex carbohydrates; in nutrition, one of the three principal classes of carbohydrates

a dietary lipid that is liquid at room temperature (e.g., olive oil, canola oil)

small intestine

urethra

starches

urinary bladder

stomach

vitamin

ureters

tube in which urine flows from the bladder to the outside of the body in both males and females; in males, the urethra also transmits semen

that portion of the digestive tract that runs between the stomach and large intestine

hollow, muscular organ that serves as a temporary, expandable storage site for the waste product urine

a complex carbohydrate that serves as the major form of carbohydrate storage in plants; starches—found in such forms as potatoes, rice, carrots, and corn—are important sources of food for animals; in human nutrition, one of the three principal classes of dietary carbohydrates, defined as a complex carbohydrate that is digestible; the other classes of dietary carbohydrates are simple sugars and fibers

a chemical compound found in foods that is needed in small amounts to facilitate a chemical reaction in the human body

an organ that performs digestion and that serves as a temporary, expandable storage site for food

two tubes in which urine is transported from the kidneys to the urinary bladder

SELF TEST

Once you have finished studying this chapter, close your books, grab a pencil, and spend the next 15 to 20 minutes completing this practice test.

Compare and Contrast

For each of the following paired terms, write a sentence of comparison ("Both") and a sentence of contrast ("However,").

fat/fibers
vitamin/mineral
mechanical digestion/chemical digestion
ureter/urethra
small intestine/large intestine

Short Answer

1. Draw and label a cross section of the digestive tract. Include the four layers of tissue.

2. Organize the following terms with respect to the production and elimination of urine: urethra, kidney, ureter, bladder.

3. What role does the liver play in digestion?

4. One of the major functions of the kidney is secretion. Explain the importance of secretion.

5. Why do you think more females than males experience urinary bladder infections? (Think about the distance a bacterium must travel from the outside of the body to the bladder in males and females.)

6. What happens to most of the water entering the digestive system each day?

7. One of the most important functions of the kidney is to maintain blood volume. How does it do this?

8. What two important processes occur in the small intestine?

9. List the six classes of nutrients, and explain why each is essential to the diet.

10. What does the stomach do?

11. Explain the functions of the accessory organs (liver, pancreas, and gallbladder) in digestion.

12. What is a food pyramid?

13. What is the relationship between the colon and the bacteria it contains?

14. What is nephron?

Multiple Choice

Circle the letter that best answers the question.

1. Which of the following terms describes the movement of water and nutrients from the kidney tubule to blood vessels?
   a. secretion
   b. absorption
   c. reabsorption
   d. concentration
   e. excretion

2. Which of the following is *not* a part of the large intestine?
   a. ileum
   b. cecum
   c. colon
   d. rectum
   e. a and b

3. Where does filtration occur within the nephron?
   a. loop of Henle
   b. Bowman's capsule
   c. collecting duct
   d. distal tubule
   e. proximal tubule

4. Which of the following describes the release of urine?
   a. secretion
   b. absorption
   c. reabsorption
   d. concentration
   e. excretion

5. Which of the following terms describes the release of water, acids, and so on, from the lining of the digestive tract?
   a. secretion
   b. absorption
   c. reabsorption
   d. concentration
   e. excretion

6. Which compartment of the digestive tract has the lowest pH?
   a. oral cavity (mouth)
   b. esophagus
   c. stomach
   d. small intestine
   e. large intestine

7. Which of the following terms describes the removal of water from the kidney filtrate?
   a. secretion
   b. absorption
   c. reabsorption
   d. concentration
   e. excretion

8. From the kidney, urine travels through _____ to _____.
   a. uretha; bladder
   b. ureters; urethra
   c. ureters; bladder
   d. bladder; ureters
   e. small intestine; bladder

9. Mechanical digestion does *not* require:
   a. tongue.
   b. teeth.
   c. saliva.
   d. food.
   e. pancreas.

10. All of the following are functions of the kidneys *except:*
    a. synthesizing vitamins.
    b. controlling body pH.
    c. maintaining water concentration in the body.
    d. retaining nutrients.
    e. excreting wastes.

11. Which of the following terms describes the movement of small organic molecules from the intestine to the capillaries?
    a. secretion
    b. absorption
    c. reabsorption
    d. concentration
    e. excretion

12. Serafina is on a new health regimen. Especially concerned about her eyesight, she has decided to take megadoses of vitamins A and E every day. You explain to her that this is not a good idea, because:
    a. vitamin K is much more important to eyesight.
    b. vitamins A and E are fat-soluble and can accumulate to toxic levels.
    c. vitamins are broken down completely in the digestive system, so they have no effect when taken as supplements.
    d. all the vitamins we need are produced within our bodies.
    e. vitamin E is a known carcinogen and should never be taken as a supplement.

13. Which of the following shows the steps of digestion in the correct order?
    a. digestion, ingestion, excretion, absorption
    b. digestion, ingestion, absorption, excretion
    c. digestion, excretion, ingestion, absorption
    d. ingestion, digestion, absorption, excretion
    e. ingestion, digestion, excretion, absorption

14. What kind of digestion occurs in the oral cavity?
    a. only mechanical
    b. only chemical
    c. both chemical and mechanical equally
    d. mostly mechanical and some chemical
    e. No digestion takes place in the oral cavity.

15. What is the function of the nervous tissue in the submucosa?
    a. to coordinate peristaltic contractions
    b. to regulate the release of digestive juices
    c. to sense the caloric content of food
    d. a and b
    e. b and c

16. What compartments are connected by the esophagus?
    a. large intestine and small intestine
    b. oral cavity and stomach
    c. stomach and pharynx
    d. small intestine and stomach
    e. pancreas and small intestine

17. What is the function of pepsin in the stomach?
    a. to digest carbohydrates
    b. to digest protein
    c. to digest lipids
    d. to increase the absorption of minerals
    e. to increase the absorption of vitamins

18. A candy bar that contains 200 Calories (150 from fat, 40 from carbohydrates, and 10 from protein) would:
    a. create enough heat to raise 1 liter of water 10 degrees Celsius.
    b. create enough heat to raise 1 liter of water 40 degrees Celsius.
    c. create enough heat to raise 1 liter of water 150 degrees Celsius.
    d. create enough heat to raise 1 liter of water 200 degrees Celsius.
    e. create enough heat to raise 1 liter of water 2,000 degrees Celsius.

19. The glycemic load is:
    a. a measure of the effect that certain foods have on blood glucose levels.
    b. a method for calculating the equivalence between fat calories and those from sugar.
    c. the relative proportions of complex carbohydrates per serving.
    d. the amount of energy required to process simple carbohydrates.
    e. a method for determining the amount of glucose required by a given type of cell.

20. Why does ingesting an alcoholic drink produce more urine than the same volume of a nonalcoholic one?
    a. Alcohol damages the glomeruli, allowing more water to leave the blood.
    b. Alcohol irritates the bladder, causing more frequent urination.
    c. Alcohol increases the absorption of water from the digestive system.
    d. Alcohol depresses the production of antidiuretic hormone.
    e. Alcohol increases the thirst mechanism, so that more water is ingested.

21. Which of the following does *not* need to occur prior to urination?
   a. relaxation of the external sphincter
   b. relaxation of the internal sphincter
   c. formation of urine
   d. activation of stretch receptors in the bladder
   e. release of antidiuretic hormone

22. When is urine actually produced?
   a. in the glomerulus
   b. in the renal pelvis
   c. in the loop of Henle
   d. in the collecting duct
   e. in the proximal tubule

23. Of the 13 vitamins required by humans, how many belong to the B complex?
   a. 12
   b. 10
   c. 8
   d. 6
   e. 4

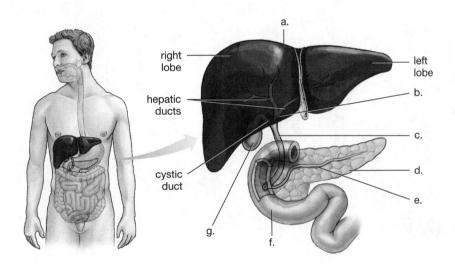

24. Match the correct term with its corresponding letter provided in the figure above.
   _____ common hepatic duct
   _____ gallbladder
   _____ liver
   _____ pancreatic duct
   _____ duodenum of small intestine
   _____ common bile duct
   _____ pancreas

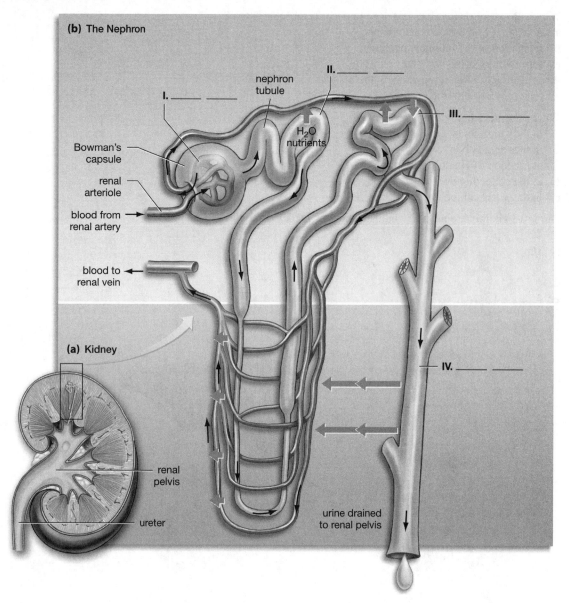

(b) The Nephron

nephron
tubule

II. ____ ____

I. ____ ____

III. ____ ____

H₂O
nutrients

Bowman's
capsule

renal
arteriole

blood from
renal artery

blood to
renal vein

(a) Kidney

IV. ____ ____

renal
pelvis

ureter

urine drained
to renal pelvis

30. Match the correct term(s) with the corresponding description, then fill in the blank spaces provided in the figure above with the correct letter and corresponding number.

a. distal tubule
b. proximal tubule
c. collecting duct
d. glomerulus

1. Blood flows in under pressure, driving some blood components out into the Bowman's capsule.

2. Water and nutrients move back into capillaries.

3. Water continues to move back into capillaries; toxins, ions, and acids move from capillaries to tubules.

4. As water continues to move back into blood circulation, waste concentrates, becoming urine that drains to the renal pelvis.

WHAT'S IT ALL ABOUT?

Here's a question to help you pull together what you've learned so far using this text.

Question: Describe the digestion of a cheeseburger with lettuce on a bun. (You might want to divide the parts of a cheeseburger into their biochemical groups—fats, proteins, etc.—and then consider their fate that way.)

What do I do now?

Remember the drill—decide what the question is asking you to do, collect your evidence from this chapter (and the others you've studied), and write!

CHAPTER 31   AN AMAZINGLY DETAILED SCRIPT: ANIMAL DEVELOPMENT

Basic Chapter Concepts

- Development proceeds from general to specific. Undifferentiated cells of the zygote differentiate into three layers of germ cells that continue to differentiate into specialized tissues and, eventually, organ systems.
- Developmental processes are controlled through a hierarchy of gene product interactions; gene products diffuse through the cells, affecting the course of development.
- The body is shaped as cells migrate to new locations, adhere to each other, and die out when necessary.

CHAPTER SUMMARY

31.1   General Processes in Development
- Development begins after fertilization, when the zygote begins to divide, forming a ball of cells of roughly the same size—an embryo.
- In the first stage of development, this zygote continues to divide without cell growth, so the cells become very small as a morula forms. The morula, which is a solid ball of cells, rearranges to produce a liquid-filled cavity surrounded by cells. This is the blastula, the defining feature of all animals. The blastula also develops a polarity; one end is referred to as the animal pole, the other, the vegetal pole.
- Gastrulation follows, in which the cells rearrange again to make three layers: an inner endoderm, a middle mesoderm, and an outer ectoderm. These cell layers contain undifferentiated cells (germ cells) that will eventually produce specialized tissues. Endoderm becomes the internal organs; mesoderm, the bone and muscle; and ectoderm, the nervous system and skin.
- The third phase of development, organogenesis, begins with the development of the nervous system. The notochord is a support structure that sends signals to the surrounding ectodermal tissue to develop into a neural tube.
- At the same time, mesodermal cells become somites on either side of the neural crest. These cells will become bones and muscle.

31.2   What Factors Underlie Development?
- Development is controlled by proteins called morphogens that activate genes to produce their products, so morphogens are transcription factors. The concentration gradient of a morphogen determines position information.
- Development is directed in a general-to-specific pattern by a hierarchical pattern of gene activation. The first genes activated specify a general position pattern. Subsequent genes fine-tune the position, and, still later, additional genes direct the development of specific structures.
- Embryonic cells are totipotent; until the eight-cell stage of development, these cells can be directed to become any cell in the body. As development progresses, the fate of groups of cells becomes predictable. Although these cells are said to be "determined," their fate may change under the influence of morphogens. Cells that are no longer susceptible to the action of morphogens are "committed"—their fate has been determined and is not changeable.

31.3   Unity in Development: Homeobox Genes
- Beyond the action of genes, the behavior of the cells of the developing embryo determines the shape of the body. Cells can move, adhere to each other, and die at the appropriate signal; these capabilities are essential to determine body shape.

31.4   Developmental Tools: Sculpting the Body
- During embryonic development, the body is shaped by three processes: cell movement; cell adhesion through proteins that protrude from the cell surface; and programmed cell death, which is highly selective.

31.5   Development through Life

- Cell death works to create spaces to create structure. Cells are not removed by other cells; rather, they self-destruct according to a predetermined program.

WORD ROOTS

blasto- = bud, budding; germ (e.g., a *blasto*cyst is an early stage in embryonic development)

gastro- = stomach (e.g., a *gastr*ula is a multilayered embryo that will give rise to tissue layers, including the stomach)

morph- = form (e.g., a *morpho*gen is a chemical that activates the formation and differentiation of tissues and organs of morphogenesis)

zygo- = yoke (e.g., a *zygo*te is a diploid cell that results from the fusion of two haploid gametes)

KEY TERMS

animal pole _____

blastocyst _____

blastula _____

cleavage _____

embryo _____

fertilization _____

gastrulation _____

induction _____

morphogen _____

morula _____

neural crest cells _____

neural tube _____

notochord _____

somite _____

transcription factor _____

vegetal pole _____

zygote _____

FLASH CARDS

To use the flash cards, tear the page from the book and cut along the dotted lines. The key term appears on one side of the flash card, and its definition appears on the opposite side.

animal pole	**fertilization**
blastocyst	**gastrulation**
cleavage	**induction**
embryo	**morphogen**

the fusion of two gametes to form a zygote; in humans (and many other organisms), the gametes known as sperm and egg fuse, resulting in a zygote

the end of a zygote with relatively less yolk and lying closer to the cell's nucleus; the location of the egg's poles defines the orientation in which the embryo develops

the process in early animal development in which an embryo's cells migrate to form three layers of tissue: the endoderm, the mesoderm, and the ectoderm; each of these layers then goes on to give rise to differing tissues and organs in the organism

hollow, fluid-filled ball of cells that is formed in the early stages of the embryonic development of humans and other mammals; in non-mammalian animals, the blastocyst is known as the *blastula*

in animals, the capacity of some embryonic cells to direct the development of other embryonic cells; for example, cells of the notochord induce the tissue above them to form the neural tube

the developmental process in which a zygote is repeatedly divided into smaller individual cells through cell division

a diffusible substance whose concentration in a region of an embryo affects development in that region

a developing organism; in humans, the developing organism from the time a zygote undergoes its first division through the end of the eighth week of development

morula

somite

neural crest cells

transcription factor

neural tube

vegetal pole

notochord

zygote

one of the blocks of mesodermal tissue in the vertebrate embryo that lies on both sides of the notochord and gives rise to muscles and the vertebrae that enclose the spinal cord

a tightly packed ball of early embryonic cells

a protein that binds to a regulatory DNA sequence, thus influencing the production of one or more other proteins

cells that break away from the top of the vertebrate neural tube as it folds together and then migrate to varying parts of the embryo, giving rise to various tissues and organs

the end of the zygote with relatively more yolk and lying farther from the cell's nucleus; the location of the egg's poles defines the orientation in which the embryo develops

a dorsal, ectodermic structure that gives rise in vertebrates to both the brain and spinal cord

a fertilized egg; in humans, the developing organism from the time of fertilization through the time of the first cell division (about 30 hours after fertilization)

a dorsal, rod-shaped support organ that exists in embryonic development in all vertebrates and in the adults of some vertebrates

SELF TEST

Once you have finished studying this chapter, close your books, grab a pencil, and spend the next 15 to 20 minutes completing this practice test.

Compare and Contrast

For each of the following paired terms, write a sentence of comparison ("Both") and a sentence of contrast ("However,").

mesoderm/ectoderm
notochord/neural tube
blastula/gastrula
determined cells/committed cells
morphogen/transcription factor

Short Answer

1. Development does not end at birth, but continues throughout the lifetime of the organism. Why?

2. Explain the process of induction.

3. How do morphogens affect development?

4. Homeobox genes are critical to the development of body form. What organism were they first discovered in, and how do they relate to human development?

5. Name and describe the three phases of development by filling in the following table:

Name of Stage	Description	Function
Cleavage		
	Cell movements and migrations	
		Formation of specialized organs

Multiple Choice

Circle the letter that best answers the question.

1. Development moves from the _____ to the _____.
   a. specific; general
   b. animal pole; mineral pole
   c. top; bottom
   d. right; left
   e. general; specific

2. A committed cell:
   a. has several options regarding its developmental fate.
   b. cannot reverse its developmental fate.
   c. is very dedicated to its job.
   d. is found in the endoderm.
   e. none of the above

3. A cell that has the potential to develop into any cell type is:
   a. committed.
   b. differentiated.
   c. totipotent.
   d. omnipresent.
   e. rapidly dividing.

4. A blastula:
   a. results from gastrulation.
   b. is a ball of cells with an internal cavity.
   c. has three germ layers.
   d. is shaped like a mulberry.
   e. all of the above

5. A gastrula:
   a. is a hollow ball of cells.
   b. results from cell movements.
   c. is shaped like a mulberry.
   d. has three germ layers.
   e. b and d

6. During development, cell A causes cells B and C to develop into specific structures. This process is known as:
   a. invagination.
   b. imbibition.
   c. inhibition.
   d. induction.
   e. incarceration.

7. Which of the following is *not* a developmental process?
   a. puberty
   b. cleavage
   c. aging
   d. cellular respiration
   e. organogenesis

8. The early embryonic structure that resembles a berry-like ball of cells is known as the:
   a. blastula.
   b. gastrula.
   c. morula.
   d. archenteron.
   e. berryball.

9. Development is controlled by the interaction of:
   a. environment and nutrition.
   b. proteins and carbohydrates.
   c. proteins and lipids.
   d. genes and proteins.
   e. cells and minerals.

10. In which of the following stages would cells be the easiest to use as a source for cloning?
    a. determined
    b. committed
    c. driven
    d. totipotent
    e. mitosis

11. Serafina needs your help. She is trying to prepare for an exam in her biology class and is having trouble with the stages of development. She shows you her study notes and asks you to correct them. Which of the following statements from Serafina's notes is a correct statement?
    a. The three layers of embryonic tissue are ectoderm, mesoderm, and wrinklederm.
    b. Organogenesis comes after gastrulation but before cleavage.
    c. Stem cells are uncommitted.
    d. The zygote stage occurs right before birth.
    e. The morula is a developmental stage in mulberries.

12. Your project in bio lab is to study the development of a bioengineered species, *Swimmerela fastus*. As you experiment with *fastus,* you realize that, similar to *Drosophila,* there is a bicoid-like substance that controls local development through changes in its concentration. Molecules that affect development in this way are known as:
    a. morphogens.
    b. antigens.
    c. pyrogens.
    d. antibodies.
    e. prions.

13. The first stage of embyronic development is _____. This process produces _____.
    a. gastrulation; a three-layered embryo
    b. gestation; a gastrula
    c. ovulation; a zygote
    d. cleavage; a hollow ball of cells
    e. parturition; a fetus

14. In the following list of development events, which occurs last?
    a. organogenesis
    b. gamete formation
    c. gastrulation
    d. cleavage
    e. spermatogenesis

15. The differentiation of an embryonic body part in response to signals from an adjacent body part is due to:
    a. contact inhibition.
    b. ooplasmic localization
    c. embyronic induction.
    d. pattern formation.
    e. all of the above

16. The process of cleavage most commonly results in a:
    a. zygote.
    b. blastula.
    c. gastrula
    d. puff.
    e. third germ layer.

17. During the early stages of embyronic development, vertebrates are generally:
    a. all the same size.
    b. all similar in shape.
    c. easy to tell apart by their distinctive shapes.
    d. easy to tell apart by the number of appendages.
    e. none of the above

18. In the fruit fly, gastrulation is responsible for dividing cells into three embryonic tissues or germ layers. Which of the following is not a germ layer?
    a. blastoderm
    b. ectoderm
    c. mesoderm
    d. endoderm
    e. gastroderm

19. A mutation in a homeobox gene could possibly result in which of the following?
    a. gastrulation
    b. physical abnormalities
    c. polyspermy
    d. cleavage
    e. none of the above

20. Programmed cell death is:
    a. an experimental technique that scientists use to kill cells.
    b. a cell suicide program crucial to normal development.
    c. a pathological condition seen only in diseased organisms.
    d. a developmental mechanism unique to earthworms.
    e. none of the above

21. Why do biologists spend such a large amount of time in understanding pattern formation in the fruit fly, *Drosophila*?
    a. Fruit flies are extremely interesting organisms.
    b. Fruit flies are easy to manipulate in the lab.
    c. Fruit flies are used as a model organism for developmental processes.
    d. Fruit flies are very complex organisms.
    e. Fruit flies are invertebrates.

22. A cell would be considered as differentiated when it:
    a. is totally committed to a particular fate.
    b. begins its pattern formation.
    c. manufactures proteins that are specific to its particular cell type.
    d. is a part of a recognizable tissue.
    e. is enclosed by a membrane.

23. Somites are:
    a. a type of muscle.
    b. a type of connective tissue.
    c. repeating blocks of tissue found on either side of the notochord.
    d. part of the spinal cord.
    e. none of the above

24. Which of the following is the tissue layer that gives rise to the brain and spinal cord?
    a. ectoderm
    b. endoderm
    c. protoderm
    d. mesoderm
    e. blastoderm

25. The bicoid protein produced in the *Drosophila* egg is a:
    a. morphogen.
    b. transcription factor.
    c. carcinogen.
    d. teratogen.
    e. all of the above

WHAT'S IT ALL ABOUT?

Here's a question to help you pull together what you've learned so far using this text.

Question: It has been said, "ontogeny (development) recapitulates phylogeny (evolution)." What does this mean?

What do I do now?
Remember the drill—decide what the question is asking you to do, collect your evidence from this chapter (and the others you've studied), and write!

CHAPTER 32   HOW THE BABY CAME TO BE: HUMAN REPRODUCTION

Basic Chapter Concepts

- Reproduction begins when eggs and sperm fuse to create a zygote in an environment favoring development.
- A human male makes gametes (sperm) continuously through his lifetime; the sperm precursor cells, spermatogonia, generate not only sperm but also new spermatogonia.
- The female makes gametes (oocytes) once a month for 35 to 40 years. All of the eggs she will ever produce are present in a suspended state of development when she is born.
- Early human development proceeds rapidly through the stages of animal development, but fetal growth occurs more slowly.

CHAPTER SUMMARY

32.1 Overview of Human Reproduction and Development

- Monthly, one oocyte matures within a follicle in the ovary under the influence of estrogen and progesterone. It is released to float down the uterine tube toward the ovary.
- In contrast, sperm are produced continuously from spermatogonia under the influence of testosterone. Sperm mature within the epididymis and travel up the vas deferens and out the urethra upon ejaculation.
- Fertilization occurs when one of the millions of sperm swimming up the vagina and into the uterus penetrates the oocyte. The surface of the oocyte changes so that no other sperm may enter. The fertilized egg continues its transit toward the uterus, ultimately implanting itself into the uterine wall.

32.2 The Female Reproductive System

- Two processes occur within the female reproductive cycle—an oocyte matures and becomes available for fertilization, and a suitable environment is prepared for the zygote.
- If the egg is not fertilized, the uterine endometrium is sloughed off, causing menstruation.
- Follicle loss occurs continuously over a woman's lifetime, so that by age 50 she may have fewer than 1,000 follicles remaining. This may be the trigger that causes the reproductive cycle to end (menopause).

32.3 The Male Reproductive System

- Sperm mature within the seminiferous tubules of the testis. About 250 million new sperm are made each day. Mature sperm move into the epididymis for additional maturation and storage; about 2.5 months are needed for the complete maturation process.
- Sperm travel through the vas deferens to reach the urethra. Sperm are mixed with seminal fluid produced by the accessory glands, seminal vesicles, bulbourethral glands, and prostate gland before ejaculation. Sperm account for only 5 percent of the volume of the semen.

32.4 The Union of Sperm and Egg

- A single spermatozoon enters the oocyte by digesting part of the outer layer surrounding the oocyte. An acrosome containing enzymes digests this layer after it becomes capacitated by the action of substances in the outer layer of the oocyte. Only one sperm succeeds in penetrating the oocyte; after penetration, the membrane potential of the oocyte changes, blocking entry of additional sperm.

32.5 Human Development Prior to Birth

- Human fetal development is completed—meaning organogenesis is complete—in 12 weeks, which is the first trimester of the pregnancy. During the second and third trimester, the fetus grows, increasing in length during the second trimester and in weight during the third.
- The human blastocyst forms two structures during early embryonic development—the embryo itself and the placenta, the organ that creates a network of maternal and fetal blood vessels to supply the fetus with nutrients.

32.6 The Birth of the Baby

- Regular contractions of the uterus at about 38 weeks post-fertilization signal the beginning of the birth process (labor). Birthing proceeds in three phases—first, the cervix must open to allow the baby to pass out of the mother; second, contractions of the uterine muscle push the baby out; and third, the placenta is expelled.

WORD ROOTS

-cyte = cell (e.g., a spermato*cyte* is a sperm cell)

oo- = egg (e.g., the *oo*gonium is a premature egg cell)

endo- = within (e.g., the *endo*metrium is the inner layer of the uterine tissue)

men- = referring to the female menstruation (*men*opause is the cessation of the monthly cycle, which normally begins with menstruation)

KEY TERMS

acrosome _____

_____

amniotic fluid _____

_____

cervix _____

_____

corpus luteum _____

_____

ectopic pregnancy _____

_____

embryo _____

_____

endometrium _____

_____

epididymis _____

_____

estrogen _____

_____

fetus _____

_____

follicle _____

_____

follicle-stimulating hormone (FSH) _____

_____

fraternal twin _____

_____

gonadotropin-releasing hormone _____

_____

identical twin _____

_____

inner cell mass _____

_____

labor _____

_____

luteinizing hormone (LH) _____

_____

menopause _____

_____

menstruation _____

_____

oocyte _____

_____

ovarian follicle _____

_____

ovulation _____

_____

placenta _____

_____

prostate gland _____

_____

semen _____

_____

seminiferous tubule _____

_____

spermatocyte _____

_____

spermatogonia _____

_____

testis _____

_____

trimester _____

_____

trophoblast _____

_____

umbilical cord _____

_____

urethra _____

_____

uterine tube _____

_____

vas deferens _____

_____

zygote _____

_____

FLASH CARDS

To use the flash cards, tear the page from the book and cut along the dotted lines. The key term appears on one side of the flash card, and its definition appears on the opposite side.

acrosome	epididymis
amniotic fluid	estrogen
cervix	fetus
corpus luteum	follicle
ectopic pregnancy	follicle-stimulating hormone (FSH)
embryo	fraternal twin
endometrium	gonadotropin-releasing hormone

a collection of tubules near the testis in which sperm complete their development and are stored

a structure located on the front end of a vertebrate sperm cell; contains enzymes that help the sperm penetrate the accessory cells surrounding the oocyte

a class of hormones, produced primarily by cells of the ovary, that supports egg development, growth of uterine lining, and development of female sex characteristics

a protective and nutritive fluid that surrounds the fetus of mammals, including filling the lungs

in humans, the developing organism from the start of the ninth week of development to the moment of birth

the lower part of the uterus, a narrow neck that opens into the vagina

in the vertebrate ovary, the complex of the oocyte (developing egg) and the cells and fluids that surround and nourish it

the structure that develops in the mammalian ovary from the ruptured tertiary follicle following ovulation; the corpus luteum secretes hormones that help prepare the reproductive tract for pregnancy

a hormone, secreted by the anterior pituitary, that promotes egg development and stimulates secretion of estrogens in women, and supports sperm production and testosterone secretion in men

an abnormal pregnancy in which the blastocyst attaches to the uterine tube or to the cervix rather than to the dorsal uterine wall

a twin who is produced first through multiple ovulations in the mother, followed by multiple fertilizations from separate sperm of the father, and then multiple implantations of the resulting embryos in the uterus

a developing organism; in humans, the developing organism from the time a zygote undergoes its first division through the end of the eighth week of development

a hormone released in tiny amounts from the brain's hypothalamus that stimulates the anterior pituitary to release two other hormones (follicle-stimulating hormone and luteinizing hormone) important in reproduction

the tissue lining the interior of the uterus in mammals, which thickens in response to progesterone secretion during ovulation and is shed during menstruation; if pregnancy occurs, this tissue houses the embryo

identical twin	ovarian follicle
inner cell mass	ovulation
labor	placenta
luteinizing hormone (LH)	prostate gland
menopause	semen
menstruation	seminiferous tubule
oocyte	spermatocyte

in human reproduction in females, the complex of an oocyte (developing, unfertilized egg) and its accessory cells and fluids

a twin who develops from a single zygote; strictly speaking, identical twins are a single organism at one point in their development and as such have exactly the same genetic makeup

the release of an oocyte from the ovary in animals

in mammalian development, the group of cells in the embryo that will develop into the baby rather than into the placenta

a complex network of maternal and embryonic blood vessels and membranes that develops in mammals in pregnancy; the placenta allows nutrients and oxygen to flow to the embryo from the mother, while allowing carbon dioxide and waste to flow from the embryo to the mother

the regular contractions of the uterine muscles that sweep over the fetus, creating pressure that opens the cervix and expels the baby and the placenta

in human males, a gland surrounding the urethra near the urinary bladder that contributes fluids to the semen

a hormone secreted by the anterior pituitary; in women, LH induces ovulation and stimulates the ovary to secrete estrogens and progestins to prepare the body for possible pregnancy; in men, LH stimulates the testes to produce androgens, such as testosterone

the mixture of sperm and glandular secretions that is ejaculated from the human male through the urethra

the cessation of the monthly ovarian cycle that occurs when women reach about 50 years of age

a convoluted tubule inside the testes where sperm development begins; immature sperm are eventually released into the interior cavity and travel to the epididymis, where sperm maturation is completed

the release of blood and the specialized uterine lining, the endometrium; occurs about once every 28 days in human females, except when the oocyte is fertilized

an immature sperm cell; spermatocytes develop from spermatogonia and develop into spermatids, which eventually develop into mature sperm

the precursor cell of eggs in the vertebrate ovary; diploid primary oocytes develop from oogonia and in turn give rise to haploid secondary oocytes in the process of meiosis

spermatogonia

urethra

testis

uterine tube

trimester

vas deferens

trophoblast

zygote

umbilical cord

tube in which urine flows from the bladder to the outside of the body in both males and females; in males, the urethra also transmits semen

diploid cells that are the starting cells in sperm production in males; spermatogonia are reproductive stem cells in that, in dividing, each of them produces one primary spermatocyte (which will develop into four mature sperm cells) and one spermatogonium

in human females, the tube that transports an ovulated egg from the ovary to the uterus, during which time fertilization may occur; also called the *Fallopian tube*

the organ in the male reproductive system in which sperm begin development and testosterone is produced

in human males, the tube that carries the sperm from the epididymis on top of the testis to the urethra for ejaculation

a period lasting about 3 months during a human pregnancy; there are three trimesters during the 9-month pregnancy

a fertilized egg; in humans, the developing organism from the time of fertilization through the time of the first cell division (about 30 hours after fertilization)

the cells at the periphery of the developing mammalian embryo that establish physical links with the mother's uterine wall and eventually develop into the fetal portion of the placenta

in human pregnancy, the tissue linking the fetus with the placenta

SELF TEST

Once you have finished studying this chapter, close your books, grab a pencil, and spend the next 15 to 20 minutes completing this practice test.

Compare and Contrast

For each of the following paired terms, write a sentence of comparison ("Both") and a sentence of contrast ("However,").

sperm/semen
uterine tube/urethra
epididymis/corpus luteum
oocyte/zygote
inner cell mass/trophoblast

Short Answer

1. What are the three stages of birth?

2. Why are the testes located outside the body?

3. What is the difference between identical and fraternal twins?

4. Why is lung function such an important factor in premature babies?

5. What is the difference between an embryo and a fetus?

6. What is a follicle?

Multiple Choice

Circle the letter that best answers the question.

1. The corpus luteum develops from:
   a. the endometrium.
   b. an immature follicle.
   c. a follicle after it has ruptured and released an oocyte.
   d. the uterus.
   e. ovarian tissue

2. What is cervical dilation?
   a. dilution of the amniotic fluid
   b. contraction of the muscles of the uterus during labor
   c. the opening of the cervix so that the baby can pass through
   d. constriction of the cervix, which holds the fetus in the uterus during development
   e. none of the above

3. Which of the following does *not* contribute to semen?
   a. kidney
   b. bulbourethral gland
   c. seminal vesicles
   d. prostate gland
   e. epididymis

4. What is polyspermy?
   a. ejaculation of multiple sperm
   b. fertilization of the oocyte by more than one sperm
   c. fertilization of more than one oocyte at the same time
   d. formation of four sperm from one precursor cell
   e. fertilization in a laboratory dish

5. A human blastocyst is equivalent to a(n):
   a. gastrula.
   b. inner cell mass.
   c. trophoblast.
   d. blastula.
   e. morula.

6. What does a fetus start to breathe during the third trimester?
   a. air
   b. blood
   c. amniotic fluid
   d. water
   e. lymph

7. How many eggs and sperm are required to produce fraternal twins?
   a. one egg, one sperm
   b. one egg, two sperm
   c. two eggs, one sperm
   d. two eggs, two sperm
   e. two eggs, no sperm

8. Ninety-five percent of semen is made up of:
   a. support material for the egg.
   b. the egg itself.
   c. sperm.
   d. support materials for the sperm.
   e. water.

9. Implantation of the blastocyst occurs:
   a. just before conception.
   b. one week after fertilization.
   c. at one month's gestation.
   d. just before birth.
   e. Implantation is not a developmental process.

10. Contractions are necessary during birth so that:
    a. the mother will know that the baby is about to arrive.
    b. the vagina will expand to accept the baby.
    c. the ovaries will know when to resume oocyte production.
    d. the cervix will expand to let the baby through.
    e. endorphins will be released, which will make the mother feel happy and calm.

11. Serafina is pregnant! She and her husband, Lee, are thrilled to be starting a family. But Serafina is worried. She hasn't mentioned it to Lee, but she hasn't felt the baby move or heard its heartbeat. You tell her not to worry; this is natural because:
   a. pregnant women often have a decrease in sensitivity of their hearing.
   b. babies rarely move before the third trimester.
   c. she is only in her first trimester, and the baby has not formed all of its structures yet.
   d. the heart does not need to start beating until birth.
   e. a thick adipose layer can often mask the baby's movements.

12. The most likely explanation for the observation that human males release so many sperm in each ejaculate is that:
   a. it requires enzymes from that many sperm to breach the fluids of the vagina.
   b. it allows men to overcome competition from other men seeking to fertilize the woman's egg.
   c. multiple sperm are required to fertilize each egg, because each sperm carries only one chromosome.
   d. the unused sperm will form the protective layer around the developing embryo.
   e. most sperm are killed by vaginal antibodies right after ejaculation.

13. During ovulation, eggs are released from the _____ into the _____.
   a. ureters; vagina
   b. urethra; oviducts
   c. ovary; oviduct
   d. oviduct; vagina
   e. ovary; vagina

14. Fertilization usually occurs in the:
   a. ovary.
   b. oviduct.
   c. uterus.
   d. cervix.
   e. vagina.

15. The function of the corpus luteum is to:
   a. secrete hormones.
   b. nourish the zygote.
   c. prevent fertilization by more than one sperm.
   d. carry the egg to the uterus for fertilization.
   e. release antibodies for the protection of the developing embryo.

16. Men don't experience menopause because:
   a. they continue to produce new gametes throughout their adult life.
   b. menopause is actually a myth; women don't experience it, either.
   c. men don't have reproductive structures that are sensitive to hormones.

   d. menopause is a monthly phenomenon, and men produce new gametes daily.
   e. they are less sensitive to hormonal changes.

17. Which of the following describes the fusion of two haploid nuclei?
   a. oogenesis
   b. atresia
   c. ovulation
   d. conception
   e. ejaculation

18. Which of the following describes the accessory structure for semen production?
   a. oviduct
   b. testes
   c. ovary
   d. urethra
   e. bulbourethral gland

19. Which of the following describes the formation of female gametes?
   a. oogenesis
   b. atresia
   c. ovulation
   d. conception
   e. bulbourethral gland

20. Which of the following describes the release of an oocyte from a follicle?
   a. oogenesis
   b. atresia
   c. ovulation
   d. conception
   e. orgasm

21. Which of the following describes the natural degeneration of follicles?
   a. oogenesis
   b. atresia
   c. ovulation
   d. conception
   e. fertilization

22. A woman normally releases one egg each month. How many sperm are released in each ejaculation from a man?
   a. 1
   b. 200,000
   c. 200,000,000
   d. 2,000,000,000
   e. 200,000,000,000

23. The endometrium is the:
   a. lining of the oviduct that sweeps the egg toward the uterus.
   b. structure that nurtures the egg before ovulation.
   c. structure that gives rise to new spermatogonia.
   d. lining of the uterus.
   e. site of sperm maturation in the testes.

24. Sperm forms in the:
  a. epididymis.
  b. seminal vesicle.
  c. prostate gland.
  d. urethra.
  e. bulbourethral gland.

25. Which of the following carries both sperm and urine?
  a. testes
  b. epididymis
  c. seminal vesicle
  d. urethra
  e. bulbourethral gland

WHAT'S IT ALL ABOUT?

Here's a question to help you pull together what you've learned so far using this text.

Question: Implantation of the embryo in the uterine wall is described as an "invasion" in the textbook; the embryo inserts itself into maternal tissue and links up with the maternal blood supply. The embryo, possessing genes from both parents, is different from the mother—it is "not self" from the mother's perspective. Why doesn't the mother's immune system attack the embryo?

What do I do now?
Remember the drill—decide what the question is asking you to do, collect your evidence from this chapter (and the others you've studied), and write!

Chapter 33 · An Interactive Living World 1: Populations in Ecology

Basic Chapter Concepts

- Ecology studies the interactions of organisms with each other and with the physical environment.
- Growth of organism populations would be exponential if it were not limited by environmental resistance.

CHAPTER SUMMARY

33.1 The Study of Ecology
- The study of ecology goes beyond environmentalism because it describes interactions that affect the living world, rather than trying to control such interactions that affect the biosphere.
- There are five scales of life that affect ecology: physiology, populations, communities, ecosystems, and the biosphere.

33.2 Populations: Size and Dynamics
- Populations can grow arithmetically or logarithmically. Arithmetical growth shows a constant rate of population increase over time, whereas logarithmic growth starts out slowly but increases very rapidly because the rate of growth is proportional to the number of individuals in the population.
- In the real world, factors in the environment such as predators, disease, and limited resources prevent logarithmic population growth.
- Most populations grow logistically; that is, when the population is small, growth rate rapidly accelerates and then slows, stabilizes, and eventually plateaus as the number of organisms in an area comes into balance with the level of available resources.

33.3 *r*-Selected and *K*-Selected Species
- The intrinsic growth rate of a species depends, in part, on whether it is a long-lived, density-dependent equilibrium species (*K*-selected) or a short-lived, density-independent, opportunistic species (*r*-selected).

33.4 Thinking about Human Populations
- Changes in population size, especially human populations, are also affected by the proportion of the population at or approaching the reproductive stage and by the movement of individuals into and out of the population.

WORD ROOTS

bio- = life; **sphere** = globe (e.g., the *biosphere* is the region of Earth and its atmosphere that are occupied by living organisms)

eco- = habitat or environment (e.g., *eco*logy is the study of the interactions between organisms and their environment)

KEY TERMS

arithmetical increase _____

_____

biosphere _____

_____

carrying capacity (*K*) _____

_____

community _____

_____

density dependent _____

_____

density independent _____

_____

ecology _____

_____

ecosystem _____

_____

environmental resistance _____

_____

exponential growth _____

_____

exponential increase _____

_____

intrinsic rate of increase (*r*) _____

_____

K-selected species _____

_____

life table _____

_____

logistic growth _____

_____

population _____

_____

r-selected species _____

_____

total fertility rate _____

_____

zero population growth _____

_____

FLASH CARDS

To use the flash cards, tear the page from the book and cut along the dotted lines. The key term appears on one side of the flash card, and its definition appears on the opposite side.

arithmetical increase	ecosystem
biosphere	environmental resistance
carrying capacity (*K*)	exponential growth
community	exponential increase
density dependent	intrinsic rate of increase (*r*)
density independent	*K*-selected species
ecology	life table

a community of living things and the physical environment with which they interact

---

an increase in numbers by an addition of a fixed number in each time period

---

all the forces in the environment that act to limit the size of a population

---

the interactive collection of all the world's ecosystems; also thought of as that portion of Earth that supports life

---

a form of population growth in which the rate of growth increases over time; exponential growth results in a J-shaped growth curve because, when plotted on a graph, the population's increase resembles the letter J

---

the maximum population density of a species that can be sustained in a given geographical area over time; in ecology, this is often denoted as K

---

an increase in numbers that is proportional to the number already in existence; this type of increase occurs in populations of living things, and it carries the potential for enormous growth of populations

---

all the populations of all species of living things that inhabit a given area; the term also is used to mean a collection of populations in a given area that potentially interact with each other

---

the rate at which a population would grow if there were no external limits on its growth; in ecology, often denoted as r

---

in ecology, effects on a population that increase or decrease in accordance with the size of that population; density-dependent effects tend to involve biological factors

---

a species that tends to be relatively long-lived, that tends to have relatively few offspring for whom it provides a good deal of care, and whose population size tends to be relatively stable, remaining at or near its environment's carrying capacity (K); also known as an *equilibrium species*

---

in ecology, effects on a population that are not related to the size of that population; density-independent effects tend to involve physical forces, such as temperature and rain

---

a table showing how likely it is for an average species member to survive a given unit of time

---

the study of the interactions that living things have with each other and with their environment

logistic growth

total fertility rate

population

zero population growth

r-selected species

the average number of children born to each woman in a human population; the most important statistic used in predicting human population changes

a form of population growth in which exponential growth slows and then stops in response to environmental resistance; also known as S-shaped growth because, when plotted on a graph, changes in the population's size resemble the letter S

state of a population in which births exactly equal deaths in a given period

all the members of a species that live in a defined geographic region at a given time

a species that tends to be relatively short-lived, that tends to produce relatively many offspring for which it provides little or no care, and whose population size tends to fluctuate widely in reaction to an environment that it experiences as highly variable; also known as an *opportunist species*

SELF TEST

Once you have finished studying this chapter, close your books, grab a pencil, and spend the next 15 to 20 minutes completing this practice test.

Compare and Contrast

For each of the following paired terms, write a sentence of comparison ("Both") and a sentence of contrast ("However,").

density dependent/density independent
r-selected species/*K*-selected species
arithmetical increase/exponential increase
community/population
exponential growth/logistic growth

Short Answer

1. Draw a population pyramid for a population past reproductive age and one for a population at or prior to reproductive age.

2. What are the three parameters involved in population dynamics?

3. In addition to birth and death rates, what factors influence human population change?

4. In the figure below, label each of these two curves as exponential or arithmetical growth. Give an example of each type of population growth.

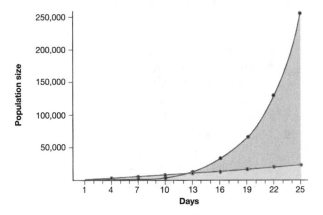

Multiple Choice

Circle the letter that best answers the question.

1. Carrying capacity is:
   a. the rate of population increase.
   b. the difference between the birth rate and the death rate.
   c. always the same for a given species.
   d. the maximum population density of a given species in a given geographical location.
   e. exponential growth of a population.

2. Immigration will tend to:
   a. stabilize population size.
   b. decrease population size.
   c. increase population size.
   d. cause a *K*-selected species to become *r*-selected.
   e. none of the above

You are doing research on the island Ecophilia. The island is very large, with some habitats and ecosystems heretofore unseen. In one particular habitat, your studies indicate that environmental conditions are relatively unstable. For example, inhabitants experience sudden torrential rainstorms followed by hot, sunny days.

3. The first species you encounter, *Mickymousiania*, is a small, mammal-like creature that seems to produce many offspring per gestation period. When their offspring are born, the parents seem to ignore them in favor of preparing for the next litter. From this you conclude that:
   a. this species is likely to be *r*-selected.
   b. this species is likely to be *K*-selected.
   c. this is an equilibrium species.
   d. this is a species with a high death rate.
   e. none of the above

4. In another habitat on Ecophilia, you discover *Dumbosiania*, a larger, less-numerous relative of *Mickymousiania*. Because it grows more slowly and has a lower reproductive rate, you conclude that the population growth of *Dumbosiania* is probably:
   a. opportunistic.
   b. *r*-selected.
   c. density dependent.
   d. density independent.
   e. none of the above

5. Which of the following is true?
   a. *K*-selection can be density independent.
   b. *r*-selection occurs in overpopulated environments.
   c. Different populations of the same species can either be *K*-selected or *r*-selected.
   d. All populations of the same species are either *K*- or *r*-selected.
   e. *r*-selected populations generally live longer.

6. In which of the following habitats would you be most likely to find the largest number of *K*-selected species?
   a. an abandoned field in Iowa
   b. the sand dunes in Kitty Hawk, North Carolina
   c. a temporary pond in the desert
   d. New Orleans following Hurricane Katrina
   e. the rain forest of Brazil

7. Populations experiencing rapid, unchecked growth would have a growth curve that resembles the letter:
   a. A.
   b. R.
   c. K.
   d. J.
   e. S.

8. A population that has stopped increasing in size because of limited resources has:
   a. become extinct.
   b. maintained its fitness.
   c. reached its carrying capacity.
   d. had to move to a new niche.
   e. little to teach us about ecology.

9. A population would grow exponentially:
   a. if it were limited only by density-dependent factors.
   b. if there were only a few predators.
   c. if there were no limiting factors.
   d. if it were a population with an equilibrated life history.
   e. if it showed logistic growth.

10. The effects of which of the following environmental factors would probably not change as a population grows?
    a. disease
    b. limited food supply
    c. competition for habitats
    d. weather
    e. predation

11. In the models that describe population growth, r stands for:
    a. population density.
    b. time intervals.
    c. total number of individuals in the population.
    d. growth rate.
    e. carrying capacity.

12. A broad-based, pyramid-shaped age structure is characteristic of a population that is:
    a. growing rapidly.
    b. at carrying capacity.
    c. stable.
    d. limited by density-dependent factors.
    e. shrinking.

13. Some populations, such as locusts, go through periods of explosive growth, followed by a decline in numbers. Their numbers are probably regulated by:
    a. predation.
    b. density-dependent factors.
    c. logistic growth.
    d. random factors.
    e. density-independent factors.

14. Chimps have a relatively low birth rate, they take good care of their young, and most live a long life. The chimp survivorship curve would look like:
    a. a line that slopes gradually upward.
    b. a relatively flat line that drops steeply at the end.
    c. a line that drops steeply at first, then flattens out.
    d. a line that slopes gradually downward.
    e. a horizontal line.

15. Of the following factors, which would have an impact on the population density in a K-selected population?
    a. disease
    b. decrease in temperature
    c. change in pH
    d. change in wind velocity
    e. global warming

16. The human population is approximately:
    a. 650,000.
    b. 6.5 million.
    c. 65 million.
    d. 6.5 billion.
    e. 65 billion.

17. A visual method of depicting data in a life table is to draw a(n):
    a. life history curve.
    b. logistic growth curve.
    c. exponential growth curve.
    d. arithmetical growth curve.
    e. survivorship curve.

18. If most of the individuals in a particular population are young, why is the population likely to grow rapidly in the future?
    a. Many individuals will begin to reproduce soon.
    b. The population will grow unevenly.
    c. Immigration and emigration can be ignored.
    d. Death rates will be low.
    e. Death rates will be high.

19. Which of the following is *not* a key event that has contributed to human population growth?
    a. development of agriculture
    b. development of medicine
    c. development of industry
    d. improvement in hygiene
    e. improvement in life quality

20. You have just completed a 3-year study of chipmunks in your local state forest. After analyzing the age structure data, you construct an age pyramid and find that it has a triangular shape. You can conclude that:
    a. the population is stable.
    b. the population is growing.
    c. the population is in decline.
    d. the population is likely to level off soon.

e. there is not enough data about the population to draw a conclusion.

21. Factors that affect the same percentage of a population regardless of the density are:
   a. density-independent factors.
   b. density-dependent factors.
   c. both density-independent and density-dependent factors.
   d. usually due to disease.
   e. usually due to predation.

22. Carrying capacity of an environment can be increased by:
   a. pollution.
   b. continuous growth.
   c. disease.
   d. population growth.
   e. natural disasters.

23. The rate of increase (r) for a population refers to which of the following relationships between birth rate and death rate?
   a. their sum
   b. their product
   c. the doubling time between them
   d. the difference between them
   e. reduction in both rates.

24. Which of the following includes all the others?
   a. ecosystem
   b. biosphere
   c. community
   d. species
   e. population

25. Populations of most species:
   a. are relatively constant over time.
   b. gradually decrease over time.
   c. gradually increase over time.
   d. vary rapidly, swinging from one extreme to the other.
   e. vary slowly over specific time periods.

WHAT'S IT ALL ABOUT?

Here's a question to help you pull together what you've learned so far using this text.

Questions: Deer have become a problem in numerous suburban areas of the United States. You have been given the responsibility of finding a way to curb the population without affecting other species in the community. Using sound ecological principles, outline how you would approach this problem, keeping in mind that there are some citizens who will resist any way to curb the population.

What do I do now?
Remember the drill—decide what the question is asking you to do, collect your evidence from this chapter (and the others you've studied), and write!

CHAPTER 34 AN INTERACTIVE LIVING WORLD 2: COMMUNITIES IN ECOLOGY

Basic Chapter Concepts

- Community ecology is the study of interactions between populations of different species.
- Species within a community interact with each other in their pursuit of resources. These interactions may be competitive or noncompetitive, and helpful, harmful, or neutral.
- Communities devastated by natural forces, or human abandonment, will regenerate in a predictable pattern of succession.

CHAPTER SUMMARY

34.1 Structure in Communities
- Loss of the keystone species significantly changes the population dynamics of the community.
- Keystone species are often the top predators within a community, but any species necessary for the dynamic equilibrium of species interactions may be a keystone species.
- Biodiversity can be measured at three levels: genetically, within communities (number of species), and across all habitats.
- Productivity is measured by how much of the sun's energy can be converted into organic structures, that is, biomass.
- Although species diversity has been shown to increase productivity, it has yet not been determined that an increase in diversity will also increase community stability.

34.2 Types of Interaction among Community Members
- Populations within a community may compete for the resources of an environment—food, shelter, and space. Because no two species can share the same resources without one species driving out the other, similar species coexist within the community only when resources can be partitioned (used at different times).
- One species may serve as a resource for another, as in the interactions defined by parasitism and predation. Predator and prey animals keep each other's populations in equilibrium because their population cycles are linked.
- Predator-prey interactions exert evolutionary pressure such that the species involved may coevolve. Prey species must develop avoidance or defense mechanisms, whereas predator species must evolve ways to overcome prey defense systems or use other species as food.
- Mutually beneficial interactions (commensalism and mutualism) are defined by the extent to which the two species depend on each other for survival.

34.3 Succession in Communities
- Communities change over time as they move through different cohorts of species. This process is known as succession.
- Primary succession occurs where some event (e.g., lava flow, human action) has removed all vegetation down to bare rock or soil.
- Secondary succession also occurs after disturbance, but in this case, some living organisms and usable soil remain.

WORD ROOTS

co- = together (e.g., *co*existence is a term used to describe the ability of two species to exist in the same environment at the same time)

KEY TERMS

Batesian mimicry _____

_____

biodiversity _____

_____

biological legacies _____

_____

climax community _____

_____

coevolution _____

_____

coexistence _____

_____

commensalism _____

_____

community _____

_____

competitive exclusion principle _____

_____

ecological dominants _____

_____

facilitation _____

_____

habitat _____

_____

host _____

_____

interspecific competition _____

_____

keystone species _____

_____

mimicry _____

_____

Müllerian mimicry _____

_____

mutualism _____

_____

niche _____

_____

parasitism _____

_____

predation _____

_____

primary succession _____

_____

resource partitioning _____

_____

secondary succession _____

_____

succession _____

_____

top predator _____

FLASH CARDS

To use the flash cards, tear the page from the book and cut along the dotted lines. The key term appears on one side of the flash card, and its definition appears on the opposite side.

Batesian mimicry	community
biodiversity	competitive exclusion principle
biological legacies	ecological dominants
climax community	facilitation
coevolution	habitat
coexistence	host
commensalism	interspecific competition

all the populations of all species of living things that inhabit a given area; the term also is used to mean a collection of populations in a given area that potentially interact with each other

a type of mimicry in which one species evolves to resemble a species that has superior protection against predators

---

when two species compete for the same limited, vital resource, one will always outcompete the other and thus bring about the latter's local extinction

variety among living things; there are three principal types of biodiversity: species diversity, geographical distribution of species populations, and genetic diversity within species populations

---

a species that is abundant and obvious in a given community; in any community, a few species, usually plants, will dominate in numbers

a living thing, or product of a living thing, that survives a major ecological disturbance; biological legacies proved to be crucial in the process of succession that occurred at Washington State's Mount Saint Helens following its eruption in 1980

---

in primary succession, the actions or qualities of earlier species that in some way assist the establishment of later species within a community

the relatively stable community that develops at the end of any process of ecological succession

---

the type of surroundings in which individuals of a species are normally found

the interdependent evolution of two or more species; coevolution can benefit both species, as in flowering plants and their animal pollinators, or it can be an arms race between species, as in a plant and its predators

---

the prey in a parasitic relationship

the condition in which two species can live in the same habitat, dividing up resources in a way that allows both to survive

---

competitive interaction between individuals of two different species

an interaction between two species in which one benefits while the other is neither harmed nor helped

keystone species	predation
mimicry	primary succession
Müllerian mimicry	resource partitioning
mutualism	secondary succession
niche	succession
parasitism	top predator

the feeding by one organism on parts or all of a second organism

a species whose absence from a community would bring about significant change in that community

in ecology, succession in which the starting state is one of little or no life and a soil that lacks nutrients

a phenomenon in which one species evolves to resemble another species

the dividing of scarce resources among species that have similar requirements; such partitioning allows species to coexist in the same habitat

a type of mimicry in which several species that have protection against predators evolve to look alike

in ecology, succession in which the final state of a habitat has been disturbed by some force, but life remains, and the soil has nutrients; a farmer's field that has been abandoned is a site of secondary succession

a form of relationship between two organisms in which both organisms benefit

in ecology, a series of replacements of community members at a given location until a stable final state is reached

a characterization of an organism's way of making a living that includes its habitat, food, and behavior

a species in a community that preys on other species but is not itself preyed on

a type of predation in which the predator gets nutrients from the prey but does not kill the prey immediately and may never kill it

SELF TEST

Once you have finished studying this chapter, close your books, grab a pencil, and spend the next 15 to 20 minutes completing this practice test.

Compare and Contrast

For each of the following paired terms, write a sentence of comparison ("Both") and a sentence of contrast ("However,").

competition/predation
primary/secondary succession
community/population
mutualism/commensalism
coexistence/coevolution

Short Answer

1. What term would you use to describe a community interaction in which one species benefits while the other experiences no change in circumstances?

2. What two processes play a role in most cases of primary succession?

3. How does predation differ from parasitism?

Multiple Choice

Circle the letter that best answers the question.

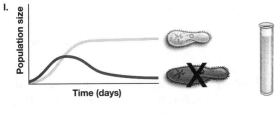

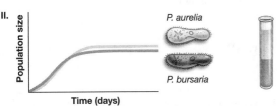

1. Figure I demonstrates _____, which is:
   a. mutualism: two populations sharing a resource equally.
   b. predation: one population devouring another.
   c. resource partitioning: two populations dividing a resource as necessary.
   d. competitive exclusion: one species driving a competitor to local extinction.
   e. parasitism: one species infecting another.

2. Figure II demonstrates:
   a. coevolution.
   b. commensalism.
   c. parastitism.
   d. resource partitioning.
   e. competitive exclusion.

3. Which of the following is true about keystone species?
   a. They are always the largest species in a community.
   b. They are always the top predator.
   c. They never interact with other species.
   d. They maintain the balance of power within the community.
   e. all of the above

4. What is a climax community?
   a. the community at the top of a mountain
   b. the community that first appears after a natural disaster
   c. the stable community achieved at the end of a succession process
   d. the community at its carrying capacity
   e. the community without its keystone species

5. A +/+ relationship, in which both populations benefit, is known as:
   a. competition.
   b. predation.
   c. mutualism.
   d. commensalism.
   e. parasitism.

6. A +/− relationship, in which one individual is harmed while one is helped, is known as:
   a. competition.
   b. predation.
   c. mutualism.
   d. commensalism.
   e. communism.

7. A −/− relationship, in which both actors may (potentially) be harmed, is known as:
   a. competition.
   b. predation.
   c. mutualism.
   d. commensalism.
   e. parasitism.

8. Communities undergoing primary succession are:
   a. stable and unchanging.
   b. not allowed to vote in some elections.
   c. found only near the equator.
   d. being engulfed by a nearby metropolis.
   e. started from bare rock.

9. When startled, a moth unfurls its wings to reveal spots that resemble owl eyes. This is an example of:
   a. commensalism.
   b. parasitism.
   c. Batesian mimicry.
   d. Müllerian mimicry.
   e. predation.

10. Which of the following could be defined as the job description of a species?
    a. population
    b. habitat
    c. niche
    d. ecosystem
    e. want ads

11. A researcher wanting to learn more about the interactions between bees and the flowers they pollinate would study:
    a. a species.
    b. a population.
    c. a community.
    d. an ecosystem.
    e. a biosphere.

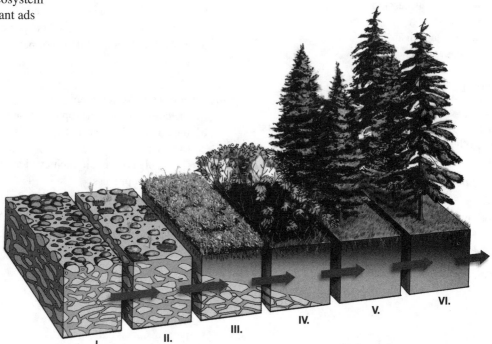

Use the figure above to answer questions 12–16.

12. A devastating fire has moved through the community depicted in the figure above, leaving few survivors. As the community goes through succession, does it begin with primary or secondary succesion? What stage is represented by your answer?
    a. primary, I
    b. primary, II
    c. secondary, I
    d. secondary, II
    e. primary, VI

13. Which of the following stages represents a climax community in the above figure?
    a. I
    b. II
    c. III
    d. V
    e. VI

14. Looking at the figure above, which stage is likely to be most stable, that is, likely to have the longest surviving cohort of species?
    a. II
    b. III
    c. IV
    d. V
    e. VI

15. Which stage has the greatest biomass?
    a. II
    b. III
    c. IV
    d. V
    e. VI

16. Compare stages III, IV, and V. Could they have occurred in a different order? Why or why not?
    a. Yes, recruitment of species is random and just happened to occur in this order on this occasion.

b. Yes, the three stages are actually simultaneous, and therefore any one of them could be a starting point.

c. No, an increase in sunlight in the habitat over time determined that species would appear in only this order.

d. No, precursor species such as those in III and IV often modify the environment and make it more conducive for succeeding populations.

e. none of the above

17. Commensalism is a ——————— relationship.

   a. +/+

   b. +/0

   c. +/−

   d. −/0

   e. −/−

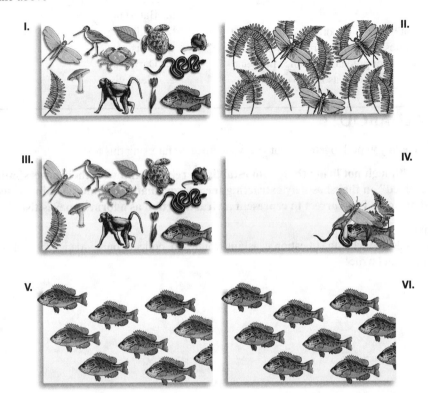

Imagine that you are doing research on the planet X-K 1157. You have been there for some months (Earth time). Your work has generated data about six habitats as depicted above. Use the figure to answer questions 18–24.

18. Which of the following exhibits low genetic biodiversity?

   a. II

   b. III

   c. IV

   d. V

   e. VI

19. In which of the following habitats did you find species evenly distributed throughout the landscape?

   a. II

   b. III

   c. IV

   d. V

   e. VI

20. In which of the following habitats did species exhibit more genetic diversity?

   a. II

   b. III

   c. IV

   d. V

   e. VI

21. Which of the following habitats had lowest species diversity?

   a. I

   b. II

   c. III

   d. IV

   e. All of these were equally diverse.

22. In one of the habitats, you found that the usable land was surrounded by a recent lava flow on which nothing could live. Which habitat was it?

   a. II

   b. III

c. IV
d. V
e. VI

23. Which habitat was a mountain lake that had been cut off from other aquatic systems for a very long time (in both Earth and X-K 1157 years)?
    a. II
    b. III
    c. IV
    d. V
    e. VI

24. On X-K 1157 you have the opportunity to chart the population dynamics of a small furry prey species, the Bo-bo, and its major predators, Madi, Mimi, and Manda. Of these, only Mimi eats Bo-bo exclusively.

Which population is most likely to cycle in conjunction with the Bo-bo population?
    a. Madi only
    b. Mimi only
    c. Manda only
    d. both Manda and Mimi
    e. both Madi and Manda

25. Uniforms of police officers are quite often blue or black. This could be cited as an example of _____ mimicry.
    a. Batesian
    b. Müllerian
    c. Gausian
    d. Darwinian
    e. Mendelian

WHAT'S IT ALL ABOUT?

Here's a question to help you pull together what you've learned so far using this text.

Question: Viruses, although not living things, do establish a relationship with their hosts similar to parasitism in that the viruses "feed" on their hosts by extracting the energy, materials, and machinery necessary to replicate and infect other hosts. Is it correct to represent a viral infection as a form of parasitism?

What do I do now?
Remember the drill—decide what the question is asking you to do, collect your evidence from this chapter (and the others you've studied), and write!

CHAPTER 35   AN INTERACTIVE LIVING WORLD 3: ECOSYSTEMS AND BIOMES

Basic Chapter Concepts

- Ecological systems are functioning units composed of living organisms and the physical (abiotic) environment.
- Energy, needed to fix carbon in biogeochemical cycling, flows through the ecosystems and is neither created nor destroyed along the way—only transformed.
- Six biomes—ecosystems of varied productivity defined by their vegetation—are determined by the climate, the average temperature, and the amount of precipitation.

CHAPTER SUMMARY

35.1   The Ecosystem
- An ecosystem is a complex structure that links together the living components (biotic factors) with the nonliving components (abiotic factors).

35.2   Abiotic Factors Are a Major Component of Any Ecosystem
- Nonliving components—resources such as minerals, nutrients, and energy sources; and conditions such as climate—are key components of every ecosystem.
- Carbon and nitrogen are key minerals that cycle between the biotic and abiotic components of the ecosystem. Carbon cycles between biotic components, mostly plants, and abiotic components, such as coal, in the form of carbon dioxide gas (CO_2). Nitrogen is also fixed by living organisms in the form of proteins and nucleic acids, but it cannot be extracted from the abiotic world—where it exists as nitrogen gas (N_2)—except by bacteria.
- Water, essential to life, cycles between large bodies of water (such as lakes, the oceans, and the polar ice caps) and water vapor in the atmosphere.

35.3   How Energy Flows through Ecosystems
- Living organisms collect and store energy temporarily; they interrupt the flow of energy from the sun to its ultimate fate—dissipation as heat.
- Producer organisms, such as plants, capture energy through photosynthesis and make it available in the form of food for consumer organisms. Producer and consumer organisms are connected by food webs known as trophic levels.
- Energy flowing from lower trophic levels (plants) to higher levels (secondary and tertiary consumers) declines by 90 percent for each step in the process. These large losses mean that organisms on the top trophic level (tertiary consumers) are rare.

35.4   Earth's Physical Environment
- The term *atmosphere* refers to all the layers of gas surrounding Earth. The bulk of the layer of gases surrounding Earth are in the trophosphere, which contains nitrogen, oxygen, carbon dioxide, and small amounts of gases such as argon and methane.
- The stratosphere contains ozone, which is important in screening out potentially harmful UV radiation. Increased use of chlorofluorocarbons (CFCs) has partially destroyed the ozone layer.

35.5   Global Warming

- The term *global warming* refers to the gradual increase in Earth's atmospheric temperature, most of which has occurred since 1950. Carbon dioxide and methane, referred to as greenhouse gases, have increased in the atmosphere and have trapped heat that comes to Earth from the sun.

35.6   Earth's Climate

- Climate affects energy flow, and climate is determined by Earth's physical environment. Atmospheric circulation acts to distribute water around the globe, creating "bands" of relatively "wet" and relatively "dry" environments around the Earth. Changes in Earth's temperature due to global warming could affect atmospheric circulation and precipitation patterns and thus change physical environments.

35.7   Earth's Biomes

- Climate—yearly temperature and rainfall patterns—determines the prevailing vegetation within a region that, in turn, determines the type of ecosystem present. These combinations of cold, warm, wet, and dry define six land biomes: tundra, taiga, deciduous forest, temperate grassland, desert, and tropical rain forest.
- The tundra and taiga represent the most severe climate conditions—cold and dry. Each area supports a variety of life forms, plants as well as small and large mammals, but species diversity is limited. Although there may be few different kinds of plants and animals, their numbers can be quite large.
- The deciduous forest and the temperate grasslands support a variety of animal and plant species, and each species is numerically well represented. These biomes support not only more types of plants and animals than are found in the tundra and taiga but also a greater number of each species.
- The tropical rain forest is the most productive biome, supporting the largest mass of organisms as well as the most diverse collection of species. This abundance of living matter—especially the trees—traps an enormous amount of carbon each year, primarily from the atmosphere. Thus the rain forest plays a critical role in decreasing the level of the greenhouse gas in the atmosphere.
- Marine ecosystems are most productive near the coast, the shallow water supporting an abundance of photosynthetic producers at the lowest trophic level. The most productive aquatic ecosystem is the coral reef, a habitat built around the accumulated carbonate skeletons of tiny sea anemones.

35.8   Life in the Water: Aquatic Ecosystems

- Ocean or marine ecosystems are becoming increasingly important in terms of their biological productivity. Most of the productivity is found near the coasts, and it is these regions that are most prone to pollution and overfishing.
- Coral reefs are the tropical rain forests of the ocean, in that they provide a habitat rich in species diversity.
- Freshwater ecosystems cover about 2.1 percent of Earth's surface and include lakes, rivers, and other running water. They can be characterized as eutrophic, meaning nutrient rich, or oligotrophic, meaning nutrient poor.
- Estuaries are regions where freshwater meets saltwater, and they are regions of high productivity.
- Wetlands, which include swamps or marshes, are critical habitat areas for migrating birds. Wetland area in the United States has been reduced by almost 55 percent since the arrival of Europeans.

WORD ROOTS

aqua- = water; **-fer** = one that bears (e.g., an *aquifer* is permeable rock that contains groundwater.)

carni- = flesh; **-vore** = eating (e.g., a *carnivore* is a flesh-eating animal)

detrit- = wear away (e.g., a *detriti*vore is an animal that feeds on dead or decaying organic material)

herb- = plant (e.g., an *herbi*vore is an animal that feeds exclusively on plants)

omni- = all (e.g., an *omni*vore is an animals that eats a variety of foods)

troph- = nutritive (e.g., a *trophic* level is a level in an ecosystem consisting of organisms that share the same sources of energy)

KEY TERMS

abiotic _____

algal bloom _____

aquifer _____

atmosphere _____

biodegradable _____

biogeochemical cycling _____

biomass _____

biome _____

biosphere _____

biotic _____

carnivore _____

CFCs _____

climate _____

coastal zone _____

consumer _____

coral reef _____

decomposer _____

desert _____

detritivore _____

ecosystem _____

element _____

energy-flow model _____

estuary _____

gross primary production _____

groundwater _____

herbivore _____

intertidal zone _____

kilocalorie _____

net primary production _____

nitrogen fixation _____

nutrients _____

omnivore _____

ozone _____

permafrost _____

phytoplankton _____

primary consumer _____

producer _____

secondary consumer _____

stratosphere _____

taiga _____

tertiary consumer _____

trophic level _____

tropical rain forest _____

_____

tropical savanna _____

_____

troposphere _____

_____

tundra _____

_____

wetlands _____

_____

zooplankton _____

_____

FLASH CARDS

To use the flash cards, tear the page from the book and cut along the dotted lines. The key term appears on one side of the flash card, and its definition appears on the opposite side.

abiotic	**biome**
algal bloom	**biosphere**
aquifer	**biotic**
atmosphere	**carnivore**
biodegradable	**CFCs**
biogeochemical cycling	**climate**
biomass	**coastal zone**

large, terrestrial regions of Earth that have similar climates and hence similar vegetative formations; at least six biome types are recognized: tundra, taiga, temperate deciduous forest, temperate grassland, desert, and tropical rain forest

pertaining to nonliving things

the interactive collection of all the world's ecosystems; also thought of as that portion of Earth that supports life

an overabundance of algae in a body of water, resulting from an excess of nutrients; the many dead algae that fall to the bottom allow decomposing bacteria to flourish, using up so much oxygen that fish can suffocate

pertaining to living things

porous underground rock in which groundwater is stored

an animal that eats meat

the layer of gases that surrounds Earth

a class of human-made chlorine compounds that destroys the atmospheric ozone that protects life on land from damaging ultraviolet radiation

capable of being broken down by living organisms

the average weather conditions, including temperature, precipitation, and wind, in a particular region

the movement of water and nutrients back and forth between biotic (living) and abiotic (nonliving) realms

the region lying between the high-tide point on shore and the point offshore where the continental shelf drops off

material produced by living things, generally measured by dry weight

consumer

energy-flow model

coral reef

estuary

decomposer

gross primary production

desert

groundwater

detritivore

herbivore

ecosystem

intertidal zone

element

kilocalorie

a conceptualization of ecosystems as units in which energy is first captured by given organisms and then transferred to other organisms

any organism that eats other organisms rather than producing its own food

an area where a river or stream flows into the ocean, bringing freshwater and saltwater habitat together; estuaries are among the most productive ecosystems on Earth

ocean structure, found in shallow, warm waters, that consists primarily of the piled-up remains of many generations of the animals called coral polyps; such reefs provide habitat for a rich diversity of marine organisms

the amount of material that a photosynthesizing organism produces through photosynthesis

a type of detritivore that, in feeding on dead or cast-off organic material, breaks it down into its inorganic components; most decomposers are fungi or bacteria

water contained within underground rock formations

a biome in which rainfall is less than 25 centimeters or 10 inches per year and water evaporation rates are high relative to rainfall

an animal that eats only plants

an organism that feeds on the remains of dead organisms or the cast-off material from living organisms

the region within the coastal zone of the ocean that extends from the ocean's low-tide mark to its high-tide mark

a community of living things and the physical environment with which they interact

the amount of energy it takes to raise 1 kilogram of water 1 degree Celsius; food consumption is measured in kilocalories, often written as *calories*

a substance that cannot be reduced to any simpler set of components through chemical processes; an element is defined by the number of protons in its nucleus

net primary production	primary consumer
nitrogen fixation	producer
nutrients	secondary consumer
omnivore	stratosphere
ozone	taiga
permafrost	tertiary consumer
phytoplankton	trophic level

any organism that eats producers (organisms that make their own food)

the amount of material a plant (or other photosynthesizing organism) accumulates through photosynthesis; net primary production is gross primary production minus energy lost to heat and the plant's expenditures of energy on its own maintenance

any organism that manufactures its own food; plants, algae, and certain bacteria are producers; by converting the sun's energy into biomass, producers capture energy that is then passed along in food webs

the conversion of atmospheric nitrogen into a form that can be taken up by living things; bacteria fix nitrogen, which is essential to life

any organism that eats a primary consumer

chemical elements that are used by living things to sustain life

the layer of Earth's atmosphere situated above the troposphere, at about 20 to 35 kilometers (13 to 21 miles) above sea level; the ozone layer lies in this level

an animal that eats both plants and animals

the biome consisting of boreal (northern) forest, characterized by cold, dry conditions, a relatively short growing season, and large expanses of coniferous trees

a gas in Earth's atmosphere consisting of three oxygen atoms bonded together that serves to protect living things from the sun's ultraviolet radiation

any organism that eats secondary consumers

the permanently frozen ground that begins about a meter below the surface in the tundra biome of the far north; neither roots nor water can penetrate this layer

a position in an ecosystem's food chain or web, with each level defined by a transfer of energy from one kind of organism to another; plants and other photosynthesizers are producers of food and thus occupy the first trophic level; organisms that consume producers are primary consumers and occupy the second trophic level, and so on

small photosynthesizing organisms that drift in the upper layers of oceans or bodies of freshwater, often forming the base of aquatic food webs

tropical rain forest

tundra

tropical savanna

wetlands

troposphere

zooplankton

a biome of Earth's far-northern latitudes characterized by very cold temperatures, very little rainfall (averaging 25 cm per year), and a short growing season

a biome found in Earth's equatorial regions characterized by warm year-round temperatures, abundant rainfall (averaging 200–450 cm per year), and great species diversity

lands that are wet for at least part of the year; wetlands are sites of great biological productivity, and they provide vital habitat for migrating birds

a grassland biome characterized by seasonal drought, small seasonal changes in the generally warm temperatures, and stands of trees that punctuate the grassland

microscopic aquatic animals that occupy a trophic level above that of phytoplankton, or photosynthesizing aquatic microorganisms

the lowest layer of Earth's atmosphere, extending from sea level to about 12 kilometers (7.4 miles) above sea level; this layer contains most of the gases in the atmosphere

SELF TEST

Once you have finished studying this chapter, close your books, grab a pencil, and spend the next 15 to 20 minutes completing this practice test.

Compare and Contrast

For each of the following paired terms, write a sentence of comparison ("Both") and a sentence of contrast ("However,").

herbivore/detritivore
gross primary production/net primary production
estuary/wetland
ecosystem/biome
abiotic/biotic

Short Answer

1. What are the two categories of abiotic factors that help to form ecosystems?

2. How do ecosystems differ from communities?

3. What is the source of the increased carbon dioxide in the present-day atmosphere?

4. What are the starting and end products of nitrogen fixation?
   a. Starting material:
   b. End product:
   c. Why is nitrogen fixation a critical process?

5. Explain the basis for the greenhouse effect.

6. What is the source of the prevailing winds found at certain latitudes of the planet, and how do these winds affect world climates?

7. What changes in the carbon cycle has the increase in industrialization brought about?

8. Place the following organisms into the appropriate boxes indicating their trophic level: blue jay, coyote, sunflower, house cat.

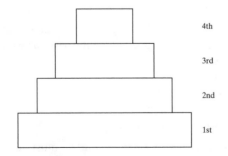

4th

3rd

2nd

1st

Multiple Choice

Match the following terms with their description. Each choice may be used once, more than once, or not at all.
   a. desertification
   b. biogeochemical cycling
   c. primary production

   d. eutrophication
   e. global warming

1. _____ Change in the profile of a lake due to increased nutrient amounts

2. _____ Formation of new desert areas

3. _____ Carbon dioxide acts as an insulator in Earth's atmosphere

4. _____ Conversion of carbon dioxide to carbohydrate

5. _____ Photosynthesis

Circle the letter that best answers the question.

6. The island of Ecophilia is unusual because its topography allows the presence of both tropical rain forest and deciduous forest biomes. Through a series of misadventures, you have been stranded on Ecophilia for several years. To survive, you must plant a garden. Why should you choose the deciduous forest as the site for your farming attempt?
   a. because that's what earlier settlers did
   b. because that will leave plenty of trees to cool the houses
   c. because the soil receives more rain there
   d. because the soil contains more nutrients there
   e. because the soil receives less sunlight there

7. Elements of Earth:
   a. are continuously created.
   b. must be recycled.
   c. are continuously replenished from space.
   d. can be obtained from deep within Earth's core.
   e. are not necessary for life.

8. Which of the following processes convert organic carbon (biomass) to inorganic carbon?
   a. photosynthesis
   b. respiration
   c. burning of fossil fuels
   d. decomposition
   e. b and d

9. Which of the following organisms can fix nitrogen?
   a. plants
   b. fish
   c. fungi
   d. bacteria
   e. all of the above

10. Based on global circulation patterns, where would you expect to find deserts?
   a. equator
   b. 30 degrees S
   c. 90 degrees N
   d. the North Pole
   e. the South Pole

11. Which biome has the highest species diversity?
    a. chaparral
    b. tundra
    c. tropical rain forest
    d. desert
    e. temperate deciduous forest

12. Which biome is characterized by a layer of permanently frozen land?
    a. tundra
    b. taiga
    c. boreal forest
    d. cold deserts
    e. none of the above

13. How are biomes defined?
    a. by the direction of the prevailing winds
    b. by the amount of groundwater
    c. by the presence or absence of nitrogen
    d. by the type of vegetation that is found there
    e. by the type of large predator that inhabits them

14. Historically, how did nitrogen enter plants?
    a. via rocks
    b. as liquid
    c. by bacteria
    d. by snake dung
    e. all of the above

15. Which of the following typifies desert biomes?
    a. hot temperatures
    b. low rainfall
    c. large trees
    d. sand dunes
    e. prairie grasses

16. Where is the most productive area of the ocean?
    a. near the shore
    b. at the equator
    c. low latitude
    d. open ocean
    e. low tide

17. When you eat an apple, you are a:
    a. primary consumer.
    b. secondary producer.
    c. primary producer.
    d. secondary consumer.
    e. tertiary consumer.

18. The main decomposers in an ecosystem are:
    a. plants and viruses.
    b. bacteria and viruses.
    c. bacteria and fungi.
    d. bacteria and plants.
    e. plants and fungi.

19. In an ecosystem, the _____ is always greater than the _____.
    a. number of producers; number of primary consumers

    b. biomass of secondary consumers; biomass of primary producers
    c. energy used by primary consumers; energy used by secondary consumers
    d. biomass of primary producers; biomass of primary consumers
    e. energy used by primary consumers; energy used by primary producers

20. Imagine that you have been shipwrecked on an island with a pair of pigs and a large supply of corn. Your best strategy would be to:
    a. feed the corn to the pigs and feed upon their offspring.
    b. kill the pigs immediately and then eat the corn.
    c. share the corn with the pigs and then eat the pigs when the corn is gone.
    d. eat only the pigs.
    e. let the pigs run freely to forage on the island.

21. The biome that is currently increasing rapidly in size is the:
    a. tundra.
    b. taiga.
    c. tropical rain forest.
    d. desert.
    e. grassland.

22. A lake in which there are too many nutrients is usually:
    a. deep.
    b. oligotrophic.
    c. eutrophic.
    d. benthic.
    e. pelagic.

23. The open ocean is less productive than coastal areas because of the limiting factor of:
    a. nutrients.
    b. temperature.
    c. oxygen.
    d. light.
    e. wind.

24. Which of the following is *not* an environmental problem that has emerged in the past 50 years?
    a. acid rain
    b. overhunting
    c. eutrophication
    d. global warming
    e. overuse of fossil fuels

25. Which of the following would be most helpful in solving the world's environmental problems?
    a. increased agricultural productivity
    b. new energy sources
    c. increased life expectancy
    d. more food from the ocean
    e. decreased human birth rates

WHAT'S IT ALL ABOUT?

Here's a question to help you pull together what you've learned so far using this text.

Question: A favorite topic of science fiction novels involves human efforts to colonize inhospitable environments on other celestial bodies by building self-contained biospheres. What kind of environmental challenges would the inhabitants of these artificial "earths" face to survive on the moon or Mars?

What do I do now?
Remember the drill—decide what the question is asking you to do, collect your evidence from this chapter (and the others you've studied), and write!

CHAPTER 36   ANIMALS AND THEIR ACTIONS: ANIMAL BEHAVIOR

Basic Chapter Concepts

- Behavioral biology seeks to determine the genetic, physiological, and environmental influences that cause animals to show specific patterns of behavior.
- Behavior can result from responses to internal cues, such as genetic predisposition, internal clocks, and hormones, or from learning actions from other members of the population.
- Social behavior, the behavior of an individual determined by interaction with other individuals of the same species, may be understood as a mechanism to increase the immediate survival of the individual as well as the long-term survival of the species' genome.

CHAPTER SUMMARY

36.1   The Field of Animal Behavior
- The study of animal behavior attempts to answer the following questions: What do animals do, why do they do it, and how do they manage to do it?
- A given behavior can be described as having a proximate or ultimate cause. Proximate causes are usually physiological or environmentally based, whereas ultimate causation rests with the species' evolutionary history.

36.2   The Web of Behavioral Influences
- An animal's behavior is usually the result of three factors: genes, learning, and environment.

36.3   Internal Influences on Behavior
- Reflexes are preset behaviors in which a specific stimulus always generates a predictable response.
- Action patterns involve a more complex response to a stimulus but also are stereotypical.
- Taxis is the movement toward (positive) or away from (negative) a stimulus. Taxis movements have a strong genetic predisposition and do not require higher reasoning ability.
- Timing of behavior relies on internal clocks. Some behaviors exhibited daily are generated by circadian rhythm, whereas behaviors such as migration can be triggered by annual clocks.

36.4   Learning and Behavior
- Learning is a combination of genetics and experience. Different types of learning require more or less reasoning ability.
- Imprinting is the matching of a preexisting genetic "frame" with an environmental "picture." During the sensitive period, an animal will add details (Tinbergen's boots) to an innate concept ("Mom").
- Habituation uses the environment to fine-tune an instinctive response; learning not to respond to a given stimulus is one example.
- Learning to link behavior with reward is known as classical conditioning. In this case, a new stimulus can be linked to an old one to generate the same behavior. This was demonstrated by Ivan Pavlov's famous dog experiments.
- Refining behavior by trial-and-error learning is known as operant conditioning.
- Insight learning requires that an animal remember the outcome of previous behaviors and be able to apply them in a novel situation. This is also described as reasoning ability.

36.5   Behavior in Action: How Birds Acquire Their Songs
- For some species of birds, song acquisition is largely genetic. However, most birds use environmental cues to learn at least part of their songs.

36.6 Social Behavior

- Animal sociality ranges from completely solitary (except for mating) to extreme eusociality, in which survival depends on the different roles adopted by group members.
- One behavior social animals use to maintain order and decrease fighting is a dominance hierarchy, also known as *pecking order.*
- When resources are limited, animals may exhibit territoriality to maintain control or access. A variety of resources may stimulate territorial behavior, from mates to food, nest sites to shelter.
- Eusocial animals (bees, wasps, ants, termites, naked mole rats) demonstrate a clear caste system in which tasks such as rearing young and gathering food are shared by only certain individuals within the group.
- Among the eusocial animals, honeybees are particularly interesting because their behavior is self-organizing.

36.7 Altruism in the Animal Kingdom

- Altruistic behavior is behavior performed for the benefit of the recipient and at some risk to the actor.
- Altruism can have an evolutionary benefit when the individuals involved are related; that is, the actor is helping to ensure the survival of copies of his or her own genome.
- In species where long-term relationships can be maintained, reciprocal altruism may be performed. In this case, one good turn generates another.

WORD ROOTS

eu- = true or actual (e.g., *eu*karyotes have a true nucleus [*karyo*])

circa- = about or around (e.g., *circa*dian means around the day [*dia*], or the time period of approximately 24 hours)

KEY TERMS

altruism _____

animal behavior _____

circadian rhythm _____

classical conditioning _____

dominance hierarchy _____

entrain _____

eusocial _____

habituation _____

imitation _____

imprinting _____

inclusive fitness _____

insight learning _____

learning _____

migration _____

navigation _____

operant conditioning _____

reciprocal altruism _____

reflex _____

releaser _____

sensitive period _____

stereotyped _____

taxis _____

territoriality _____

FLASH CARDS

To use the flash cards, tear the page from the book and cut along the dotted lines. The key term appears on one side of the flash card, and its definition appears on the opposite side.

altruism

habituation

animal behavior

imitation

circadian rhythm

imprinting

classical conditioning

inclusive fitness

dominance hierarchy

insight learning

entrain

learning

eusocial

migration

a simple form of learning, consisting of a reduction in a response, based on repeated exposure to a stimulus that has no positive or negative consequences

a costly or risky behavior carried out by one animal for the benefit of another animal

a sophisticated form of learning in which one animal copies behavior it observes in another animal

a subdiscipline of biology concerned with the study of the behavior of animals

a process of learning that results in one animal preferentially associating with another

biological cycles that can function independently of environmental cues and that are roughly synchronized to Earth's 24-hour rotation

an individual's relative genetic contribution to a succeeding generation, made both through itself and through relatives who have reproduced because of assistance the individual has provided

a form of learning in which animals learn to respond in a customary way to a new stimulus that has been paired with an existing stimulus; in the case of Pavlov's dog, a customary response (salivation) was elicited when a new stimulus (the sound of a bell) was paired with an existing stimulus (food delivery)

a sophisticated form of learning in which an animal makes associations between objects or events that it has previously regarded as unrelated; in short, the ability to reason

a persistent power ranking in an animal population that gives those of higher rank the ability to control some aspect of the behavior of those of lower rank

the acquisition of knowledge through experience

the initiation of a new cycle of an organism's internal clock; many biological rhythms are entrained by such environmental cues as sunlight and temperature

a regular movement of animals from one location to a distant location; also, the movement of individuals from one population into the territory of another population; migration is the basis of gene flow among populations

a species of animal that is organized into a caste system in which there is a division of labor, such that different members of a population will consistently perform different tasks, and in which the young are raised through cooperative care, meaning care provided by many members of the group

navigation

operant conditioning

reciprocal altruism

reflex

releaser

sensitive period

stereotyped

taxis

territoriality

the time interval during which an animal can learn to respond to a given stimulus

in animals, the use of various cues to enable movement toward a desired location

performed precisely the same way each time by individual animals of the same species

a complex form of learning that occurs when experience teaches animals to associate one of their own actions with a particular outcome; such learning occurs when an animal's own behavior brings about a response that has either negative or positive consequences

in animals, a genetically predisposed movement toward or away from a stimulus; a positive taxis is a movement toward the stimulus; a negative taxis is a movement away from the stimulus

an exchange of altruistic acts by individual animals over time

a phenomenon in which animals try to keep other animals out of a given area; in general, territoriality refers to efforts to keep members of an animal's own species from entering an area

automatic nervous system response that helps an organism avoid danger or preserve a stable physical state; the knee-jerk response is a well-known reflex

the critical element in an action or object that triggers an action pattern in animals; a releaser can be visual, such as a color, but odors, touch sensations, tastes, and sounds can also trigger behavior

SELF TEST

Once you have finished studying this chapter, close your books, grab a pencil, and spend the next 15 to 20 minutes completing this practice test.

Compare and Contrast

For each of the following paired terms, write a sentence of comparison ("Both") and a sentence of contrast ("However,").

navigation/migration
proximate/ultimate causes of behavior
taxis/reflex
habituation/imprinting
classical conditioning/operant conditioning

Short Answer

1. What is the relationship between natural selection and the causation of behavior?

2. What is the diffence between internal and external influences on behavior?

3. Give an example of a behavioral "action pattern." Explain the significance of this type of behavior.

4. Give an example of a hormone influencing a behavior in males.

5. How does altruism differ from reciprocal altruism?

6. Describe some the costs and benefits of being a social species.

7. Explain how song acquisition in birds could involve both learning and internal influences.

Multiple Choice

Circle the letter that best answers the question.

1. Which of the following does *not* play a role in influencing animal behavior?
   a. natural selection
   b. genetic predisposition
   c. environmental influence
   d. learning
   e. All of these play roles in influencing animal behavior.

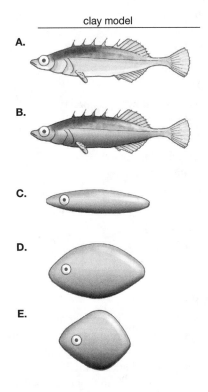

clay model

A.

B.

C.

D.

E.

Use the figure above to answer questions 2–4.

2. In the figure above, the model that failed to cause a threat posture in the experimental subject was:
   a. A.
   b. B.
   c. C.
   d. D.
   e. E.

3. Your answer to question 2 was based on the fact that the "nonthreatening" model:
   a. had no spines.
   b. was too tall.
   c. was too flat.
   d. was not red.
   e. did not move.

4. The work illustrated in the figure above was conducted by:
   a. Konrad Lorenz.
   b. Karl von Frisch.
   c. Niko Tinbergen.
   d. Ivan Pavlov.
   e. Sigmund Freud.

5. Movement toward or away from a stimulus is properly known as:
   a. migration.
   b. navigation.
   c. taxis.
   d. circadiation.
   e. rhythmicity.

6. Which of the following is *not* a type of learning?
   a. imprinting
   b. habituation
   c. classical conditioning
   d. operant conditioning
   e. reflex

7. Insight learning can be described more simply as:
   a. inborn knowledge.
   b. reasoning ability.
   c. instinctual behavior.
   d. habituation.
   e. imprinting.

8. Suppose that you are asked to move home and help raise your siblings. To do so, you would have to give up on a promising relationship that could result in your only chance at marriage and children. Based on your understanding of kin selection and inclusive fitness, at least how many siblings would it take to equal your potential contribution to future generations?
   a. 1
   b. 2
   c. 4
   d. 8
   e. 16

9. Your dog accidentally catches a paw in the refrigerator door handle while chasing a fly. As she pulls her paw free, the door opens, and she discovers the repository of all those great leftovers you normally don't let her eat. Gradually, she learns to pull open the door and grab a snack whenever you leave the room. This type of learning is known as:
   a. imprinting.
   b. habituation.
   c. classical conditioning.
   d. operant conditioning.
   e. altruism.

10. You and your friend John-Karl each adopt a duck egg being hatched in an incubator. You spend a lot of time with your duckling during the first few hours after hatching, whereas John-Karl leaves his duckling in a box with plenty of food and water and goes bowling. The next day, your duckling tries to follow you everywhere, but John-Karl's just wanders aimlessly. John-Karl can't understand the different behavior of the two until you explain.

What key element did he miss in the process of imprinting his duckling?
   a. the habituation hour
   b. the dominance hierarchy
   c. the duckling adoption call
   d. the trial-and-error period
   e. the sensitive period

11. Hormones play a role in the _____ cause of a given behavior.
   a. genetic
   b. environmental
   c. reactionary
   d. proximate
   e. ultimate

12. You are returning to the island of Ecophilia to do more research. As you set up your study, you make a list of questions that you feel are essential to your understanding of two species, *B. andrewsiana* and *B. mikosiana*. Which of the following is *not* a question fundamental to describing the behvioral biology of this species?
   a. What do *B. andrewsiana* and *B. mikosiana* do?
   b. Why do *B. andrewsiana* and *B. mikosiana* perform certain behaviors?
   c. What is the evolutionary significance of these behaviors?
   d. Where do *B. andrewsiana* and *B. mikosiana* perform these behaviors?
   e. How do *B. andrewsiana* and *B. mikosiana* know how to perform these behaviors?

13. Which of the following requires the greatest level of learning (and the least amount of genetic programming)?
   a. operant conditioning
   b. habituation
   c. insight
   d. reflex
   e. imitation

14. Which of the following would be required for migratory behvaior?
   a. circadian rhythm
   b. instinctual rhythm
   c. taxis
   d. annual clock
   e. imprinting

15. What is the ultimate cause of behavior?
   a. natural selection
   b. imprinting
   c. genetics
   d. learning
   e. habituation

16. One day, with some time to spare on your hands, you sit by your window and watch birds forage for seeds in the yard outside. You notice that different

individuals take turns acting as sentinels. That is, instead of feeding, they spend their time looking around, as if in search of predators. You call your roommate over to watch as well. When your roomie asks what all the excitement is about, you explain that what you are seeing is an example of:

a. habituation.
b. taxis.
c. altruism.
d. territoriality.
e. mating behavior.

17. My older sister marries my husband's brother, whereas my younger sister marries a close friend of the family. We all have children. When I discover that my daughter needs a bone marrow transplant, I appeal to my sisters and their families to be tested as possible donors. Who in this family group is most likely to be a suitable genetic match for my daughter?

a. my older sister
b. my older sister's children
c. my younger sister's children
d. my older sister's husband
e. my younger sister

18. Given that, if the situation were reversed, I would do the same for my sisters' children, what kind of behavior am I asking them to demonstrate?

a. insight learning
b. nonreciprocal altruism
c. reciprocal altruism
d. territoriality
e. eusociality

19. When your cat Fluffy was a kitten, she delighted in jumping onto the table as you sat down to eat. At first you just lifted her off the table, but the behavior continued until you began enclosing Fluffy in her travel box, which she loathes, for 10 minutes after her table-top excursion. After four such imprisonments, Fluffy stopped interrupting your dinner. Why?

a. habituation
b. operant conditioning
c. classical conditioning
d. altruism
e. reflex

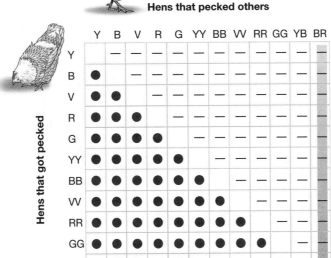

Hens that pecked others

Use the figure above to answer questions 20–22.

20. The figure above is an example of:
a. territoriality.
b. dominance hierarchies.
c. altruism.
d. operant conditioning.
e. classical conditioning.

21. In the figure above, _____ always wins.
a. Y
b. R
c. GG

d. YB
e. BR

22. In the figure above, _____ wins slightly more often than she loses.
a. Y
b. R
c. GG
d. YY
e. BR

23. Eusociality has been demonstrated among all of the following *except:*
    a. humans.
    b. ants.
    c. termites.
    d. bees.
    e. naked mole rats.

24. Which of the following can cause an organism to demonstrate territoriality?
    a. food
    b. nesting space
    c. mating partners

d. limited shelter
e. All of these can stimulate territorial behavior.

25. Your pet bird wakes you every morning by squawking and chirping just as the sun begins to rise. Because your bird performs this behavior at the same time each day, you could describe Birdie's behavior as an example of:
    a. annual clock.
    b. circadian rhythm.
    c. migratory behavior.
    d. territoriality.
    e. sociality.

WHAT'S IT ALL ABOUT?

Here's a question to help you pull together what you've learned so far using this text.

Question: Of all of the behaviors in the animal world, altruism is the most puzzling, especially when one considers the human animal. One could explain the altruistic behavior of parents for their children as merely the effect of demands of inclusive fitness. However, inclusive fitness predicts that parents should stop feeling an attachment to their children once their offspring are independent, yet this is rarely the case (trust us: we're both daughters and mothers, so we know). What is the value of the continued altruism shown by parents for their children?

What do I do now?

Remember the drill—decide what the question is asking you to do, collect your evidence from this chapter (and the others you've studied), and write!

ANSWER KEY

CHAPTER 1 Science as a Way of Learning: A Guide to the Natural World

Compare and Contrast

variable/constant

Both refer to parts of a scientific investigation. However, a variable is something that can change throughout the investigation, whereas a constant is kept unchanged throughout the investigation.

theory/hypothesis

Both are important components of the process of science and problem solving. However, a theory is a set of principles that is supported by evidence, whereas a hypothesis is a probable explanation for a set of observations.

atoms/molecules

Both refer to levels of organization below the level of the cell. However, atoms are the fundamental units of matter, whereas molecules are atoms that are chemically bonded to one another.

cell/organisms

Both are levels of organization of life. However, the cell is the building block of life, whereas organisms may be unicellular or multicellular individuals.

population/community

Both refer to collections of organisms. However, a population is a group of organisms in a geographic area, whereas a community is a collection of populations of different species in a geographic area.

Short Answer

1. Prior to 1850, there were few microscopes available and thus people had very little idea of the microorganisms that covered all the earth. Therefore, when maggots were seen on food that had been out for several days, it would seem that the worms were created spontaneously.

2. A hypothesis is used to design experiments that will yield facts. These facts can then be used to build a theory.

3. *Falsifiable* means that the hypothesis (or explanation) can be disputed. The principle of falsifiability allows scientists to discard ideas that are not supported by facts.

4. Because any single feature of a living organism may be replicated in an inanimate object.

Multiple Choice

1. d
2. b
3. e
4. e
5. d
6. b

7. a
8. c
9. e
10. b

11. e
12. e
13. b
14. e

15. e

CHAPTER 2 Fundamental Building Blocks: Chemistry, Water, and pH

Compare and Contrast

polar/nonpolar

Both involve the joining of atoms to form compounds or molecules through the sharing of electrons. However, non-polar bonds involve the equal sharing of electrons, whereas polar bonds have unequally shared electrons.

inert/reactive

Both refer to atoms. However, reactive atoms have unfilled valence shells, whereas inert atoms have filled valence shells.

covalent/ionic

Both are types of bonds. However, covalent bonds result from the sharing of electrons, whereas ionic bonds are the attractions between oppositely charged ions.

element/matter

Both terms refer to material comprising the universe. However, elements are pure forms of matter.

solute/solvent

Both terms refer to the materials that comprise a solution. However, solute is solid and solvent would be liquid.

Short Answer

1. The nucleus contains protons and neutrons. The electrons are found in shells, or orbitals, around the nucleus.
2. The atom would have fewer electrons because it is neutral.
3. A three-dimensional shape determines how an atom or molecule can fit together with others. Therefore, shape determines its ability to function.
4. With less water in the atmosphere, Arizona will likely experience more variation during the day.
5. If you test the solution for pH and the value is below 7, it is an acid. A value above 7 indicates a base.

Multiple Choice

1. b	8. c	15. c	22. c
2. c	9. b	16. a	23. e
3. a	10. c	17. b	24. c
4. a	11. c	18. e	25. a
5. b	12. e	19. d	
6. d	13. b	20. a	
7. c	14. c	21. c	

What's It All About?

1. **Type of question:** Already defined.

2. **Collect the evidence:**

Molecular Bonds	Molecular Bonds
• Type of bond made by atoms depends on atomic structure. Bonds made put atoms in most stable state.	• Solubility: from Chapter 1—"like dissolves with like"; molecules with polar bonds will dissolve in polar solvents.

Molecular Bonds	Molecular Bonds
• Covalent: sharing electrons between atoms; shared electrons fill up outer electron shells of participating atoms; can be polar (if electrons not shared equally—example, water) or nonpolar.	• Shape: two-dimensional; bonding determines how atoms line up in molecules, whether or not the molecule will have straight (H_2) or bent (H_2O).
• Ionic: electrons are gained or lost; creates ions held together by electrostatic attractions—example, NaCl.	• Shape: three-dimensional; determines molecules' activity, what it binds to—example, signal molecules and receptors.
• Hydrogen bonding: always involves hydrogen (especially water's hydrogen atoms) pairing with an electronegative oxygen or nitrogen. Relatively weak bonds.	

3. **Pull it all together:**

Bonds create stable structures. Atoms bond to each other, creating molecules, because making a bond puts the atoms into their most stable state. Atoms achieve this stable state by sharing electrons, as in a covalent bond, or by losing or gaining electrons, as in an ionic bond. In both cases, the type of bond affects the solubility and shape of a molecule.

The solubility of a molecule is a function of its type of bond. Atoms can form covalent bonds by sharing electrons and filling up their outer electron shells. If the electrons are shared equally, as in methane, the molecule will be nonpolar. When the electrons are shared unequally, as in water, the molecule will be nonpolar. A water molecule is nonpolar because the oxygen atom pulls the shared electrons closer to its nucleus, gaining a slight negative charge, whereas the hydrogen atoms, in which the electrons are farther away, have a slight positive charge. Polar molecules can dissolve in polar solvents; nonpolar molecules prefer nonpolar solvents. Thus, the nature of the covalent bond will determine whether the molecule is water-soluble (polar solvent) or not.

If one atom gives up an electron to a second atom, ions are created. Ions are charged atoms: positively charged if an electron is lost, or negatively charged if an electron is gained. Because the positive and negative ions are electrostatically attracted to each other, they remain held together by an ionic bond. An example of a molecule held by an ionic bond is table salt, NaCl, and because both the sodium and the chloride are charged, salt is soluble in water.

Another property affected by bonding is the three-dimensional shape of a molecule. Both covalent and ionic bonds can produce molecules that are linear or have more elaborate shapes, such as pyramids or cubes. The nature of the bond, polar or nonpolar, also determines how individual molecules interact with each other. Water molecules, for example, can form weak bonds (hydrogen bonds) between each other because of the partial charges on the atoms. Hydrogen atoms in larger molecules may also participate in hydrogen bonding and determine the three-dimensional shape of the molecule.

CHAPTER 3 Life's Components: Biological Molecules

Compare and Contrast

cellulose/starch

Both are complex carbohydrates found in plants. However, cellulose is a rigid molecule found in plant cell walls, whereas starch is the main form of energy storage in plants.

polysaccharide/polypeptide

Both are polymers made of many smaller monomer subunits. However, a polysaccharide is a carbohydrate molecule, whereas a polypeptide is a protein molecule.

fatty acid/triglyceride

Both are types of lipids. However, a fatty acid is a building block of a lipid, whereas a triglyceride is a lipid made of fatty acids and another molecule, glycerol.

DNA/RNA

Both are nucleic acids and store information. However, DNA is a double helix and is found in the nucleus, whereas RNA is a single helix that can be found both in the nucleus and the cytoplasm.

lipoprotein/glycoprotein

Both are complex molecules containing proteins. However, a lipoprotein is a combination of a lipid and a protein, whereas a glycoprotein is a combination of a carbohydrate and a protein.

Short Answer

1. Carbon is a common element with the ability to form up to four separate bonds. This property allows it to act as a "universal connector."

2. Nucleic acids code for proteins and aid in protein synthesis.

3. A triglyceride would contain glycerol (a 3 C molecule) plus three fatty acid molecules; see Figure 3.9 in Chapter 3. The condensation reactions would occur between the H on the glycerol and the OH on the fatty acid molecules. The by-product is three water molecules.

4. Anything containing glucose or fructose would provide you with immediate energy because these molecules are monosaccharides and would be metabolized quickly in your body.

5. Two conditions that might affect the structure of egg whites are (1) beating egg whites, as in making meringue; and (2) heating egg whites, as in cooking an egg. Both denature the protein by changing its structure.

6. Both the amino (NH_2) and carboxyl (COOH) groups are found in every one of the 20 amino acids. In order to make a protein, at least two amino acids would be joined together in a dehydration reaction, which would result in a water molecule as a by-product.

Multiple Choice

1. e	8. a	15. b	22. b
2. b	9. c	16. a	23. b
3. e	10. a	17. b	24. c
4. e	11. c	18. c	25. a
5. c	12. d	19. c	
6. d	13. c	20. b	
7. b	14. c	21. b	

What's It All About?

1. **Type of question:** Already defined.

2. **Collect the evidence:** Advantages and examples are as follows.
   a. Ease of construction—polymers can assemble with one type of bond.

Examples:

- All simple sugars join to make carbohydrates the same way, splitting out water.
- Amino acids join to make proteins—different proteins have different orders of amino acids.
- Nucleic acids are chains of nucleotides.
- Lipids don't follow a monomer-to-polymer pattern, but different classes of lipids share similar structures.

Insight #1—If the bonds making a certain polymer are always the same, that means that the cellular machinery for making the bond is always the same. It's as if everything is assembled with one size of screw, so the only tool we need is a screwdriver.

Insight #2—If we need only one "tool," we have an efficient assembly line that probably works pretty fast.

   b. Greater diversity of molecules—arranging the order of monomers in different ways results in different shapes and thus different molecules.

Examples:

- Carbohydrates—linear molecules are structural; storage forms have branches.
- Proteins—amino acid chain directs folding and determines the shape of the protein.
- From Chapter 2, we know that shape also affects molecular interactions.
- Nucleic acids—the order of nucleotide bases contains information that directs protein synthesis.

Insight—Because a molecule's activity depends on its shape (as we learned in Chapter 2), different shapes mean different functions. Thus, when we assemble monomers in different orders we should get different shapes, which means that from a common pool of monomers, we can make a large number of molecules with many different functions.

3. **Pull it all together:**

Creating polymers from monomers offers two advantages to the organism: ease of molecular construction and increased diversity of molecules. Carbohydrates, proteins, and nucleic acids are each polymers of monomeric units that have essentially identical structures. Within each class of molecules, each monomer has common features; therefore, the bonds joining the monomers are identical. The simple sugars that make up more-complex carbohydrates are all linked by a C–O–C bond. Proteins are made when the C on one end of an amino acid joins with the N of another. This means that the cell needs only one type of "tool" to assemble all carbohydrate polymers, and one "tool" for all of the proteins. This implies that polymers can be made quickly and efficiently.

In addition, the cell can make many different molecules from the same collection of monomers just by changing the order of the monomers. For example, by connecting the amino acids in different orders, the cell can produce many different proteins with many different shapes from the same collection of amino acids. Because the function of a protein depends on its shape, each of these molecules can have a unique function. Many carbohydrates are polymers of the same sugar monomer, so the function of the polymer depends on where the cell connects the monomers. Some carbohydrates, such as cellulose, are linear molecules, a characteristic that makes them rigid and very strong. By making branches on the chain of sugar monomers, the cell makes a polymer, such as starch, that is easy to store.

CHAPTER 4 Life's Home: The Cell

Compare and Contrast

cilia/flagella

Both are microtubule-containing structures. However, cilia are shorter than flagella, and there are more cilia per cell than flagella.

nucleus/nucleolus

Both are not part of the cytoplasm. However, the nucleolus is found within the nucleus.

cytosol/cytoplasm

Both are outside the nucleus. However, only cytoplasm contains the organelles.

smooth ER/rough ER

Both are membrane-enclosed lumens. However, the rough ER has associated ribosomes.

prokaryote/eukaryote

Both are types of living organisms. However, eukaryotes have membrane-bound organelles, but prokaryotes do not.

Short Answer

1. Plants have plasmodesmata, and animals have gap junctions for communication. These structures provide an opportunity for coordination and control in multicellular organisms.

2. Plants rely on a rigid cell wall and a filled central vacuole for support. In animals, an exoskeleton or endoskeleton carries out this function.

3. The mitochondria is like the factory power plant.

4. The proteins would probably be degraded and recycled.

5. See Figure 4.5.

6. The nucleus contains the genome, the DNA that codes for cellular structures. It also contains accessory material (protein, RNA) involved in nucleuar processes.

7. Prokaryotes only have one kind of organelle, ribosomes.

8. The presence of mitochondria allows eukaryotic cells to increase their energy-harvesting efficiency by 1,500 percent. The mitochondrion is the site of aerobic respiration.

9. Lysosomes contain enzymes that are important in digesting and recycling cellular material.

Multiple Choice

1. d	8. a	15. d	22. b
2. d	9. b	16. a	23. a
3. c	10. a	17. c	24. c
4. c	11. b	18. a	25. c
5. d	12. a	19. d	
6. c	13. d	20. c	
7. a	14. a	21. a	

Labeling Exercise

Refer to Figure 4.4 in your main textbook for answers.

What's It All About?

1. **Type of question:** Already defined.

2. **Collect the evidence:**
   a. Molecular composition of cellular structures—in general, cellular structures are comprised of phospholipids or proteins.
   b. Function of cellular structures—consult the text.

3. **Pull it all together:**
   All parts of the cell are made from the four basic types of molecules: lipids, proteins, nucleic acids, and carbohydrates. Each class of biomolecule contributes to the specialized functions of the cell's component parts. Lipids, specifically phospholipids, make up cellular membranes, so every organelle in the cell, such as mitochondria and the ER, depends on a lipid membrane for its existence. Without membranes, cells would not exist. The cell's internal skeleton is composed of proteins, as are all of the enzymes that carry out cellular functions. The protein actin makes up the cytoskeletal fibers that give a cell its shape. Enzyme proteins carry out many different jobs, such as making energy inside the mitochondria and recycling biomolecules inside the lysosome. The carbohydrate cellulose makes plant cell walls rigid and strong. The nucleic acid DNA stores information in the cell nucleus. This information is copied to RNA, another nucleic acid, and is used to direct synthesis of proteins on ribosomes, which are also made of RNA.

CHAPTER 5 Life's Border: The Plasma Membrane

Compare and Contrast

phospholipids/cholesterol

Both are important components of cell membranes. However, phospholipids form the major portion of the membrane, whereas cholesterol helps maintain its fluidity.

passive diffusion/facilitated diffusion

Both involve the movement of material down the concentration gradient across a semipermeable membrane. However, facilitated diffusion involves the use of a transport protein.

phagocytosis/pinocytosis

Both are methods of moving material into a cell. However, phagocytosis ("cell eating") requires the use of pseudopodia to engulf material.

integral proteins/peripheral proteins

Both are proteins that are associated with the cell membrane. However, integral proteins are bound to the hydrophobic interior of the membrane, whereas peripheral proteins are not bound in this way.

hypertonic/hypotonic

Both terms refer to conditions in which a cell can exist relative to osmosis. However, a cell that is in a hypertonic solution has is a greater concentration of solutes outside the cell relative to inside the cell, whereas a cell that is in a hypotonic solution has a lesser concentration of solutes outside the cell.

Short Answer

1. The part of the membrane that transports chloride is defective; as a result, the chloride concentration is greater inside the cells than outside, and water moves into the cells because of osmosis. This leaves less water outside the cells to dilute the mucus, so the mucus is stickier than normal.
2. Proteins carry out functions of structural support, recognition, communication, and transport.
3. Osmosis would likely cause water to move from the surroundings into the cells, leading to swelling. The extreme outcome might be lysis of the cells, because animal cells have no cell walls.

4.

Substance	Cross solo?	Reason
Water	Yes	It is small; but because it is charged, it would move through faster with a protein.
Oxygen	Yes	Gases freely diffuse.
A large protein	No	It is too large.
A small molecule with a large charge	No	The charge will not allow transit through the lipid bilayer.
A small molecule with a large charge	Yes	Same reason as water.

Multiple Choice

1. d	8. c	15. a	22. a
2. c	9. c	16. c	23. b
3. a	10. c	17. a	24. c
4. d	11. a	18. c	25. c
5. b	12. c	19. d	
6. d	13. a	20. e	
7. d	14. d	21. b	

What's It All about?

1. **Type of question:** Already provided.
2. **Collect the evidence:**
   Your collection of examples need not be exhaustive, just convincing, so look for diverse examples of how membranes make it possible for organisms to do what they do. Use the information in Chapter 5 to answer this question.

Features of Living Things	Role of Plasma Membrane(s) (PM)
Assimilate and use energy	Transport—PM helps move fuel molecules into the cell, using specialized transport structures. Can concentrate molecules in cell.
Respond to environment	Communication—Receptor proteins and the glycocalyx transmit information about the outside world to inside the cell.
Maintain constant internal environment	Structure—Phospholipids create a barrier, keeping useful molecules inside and harmful ones outside. Also, transport function controls entry and exit of molecules.
Highly organized	Has shape determined by lipid bilayer; contains protein specific to type of cell.
Made of cells	No membrane = no cell.

3. **Pull it all together:**
   Because plasma membranes allow living organisms to do the things they do, we can argue that the simplest definition of a living thing is something that has plasma membranes. All living things are made of cells, and cells cannot exist without a plasma membrane to separate what is inside the cell from the environment. Membranes are highly organized structures: they have a shape because of the nature of the phospholipid bilayer, and this bilayer contains proteins that identify the type of cell. The plasma membrane creates a barrier, keeping useful molecules on the inside and useless and harmful ones outside, so that whatever is inside a cell will stay there until the cell decides to move it out. The receptor proteins and glycocalyx embedded in the membrane communicate information about the outside world to the inside of the cell so that the cell can respond to its environment. The special transport proteins in the plasma membrane allow the cell to concentrate molecules on the inside that the cell can use to make energy. Thus, the plasma membrane makes it possible to accomplish many of the characteristics of living organisms.

CHAPTER 6 Life's Mainspring: An Introduction to Energy

Compare and Contrast

endergonic/exergonic

Both describe energy flow in a chemical reaction. However, they describe an opposite energy difference between the reactants and products of a reaction.

enzyme/coenzyme

Both work to catalyze chemical reactions. However, a coenzyme helps an enzyme.

potential energy/kinetic energy

Both are forms of energy. However, potential energy is stored energy, whereas kinetic energy is the energy of motion.

competitive inhibitor/allosteric inhibitor

Both inhibit enzyme activity. However, they do so by binding at different sites on the enzyme.

metabolic pathway/metabolism

Both describe the chemical reactions that take place within the cell. However, a single metabolic pathway would be a subset of the overall cell metabolism.

Short Answer

1. Energy transformations (potential to kinetic) are leaky, and much energy is lost as heat. Therefore, we must continue to supply our bodies with energy sources (food).

2. Kinetic energy is the energy used to actually perform some action.

3. Using energy from one set of reactions, cells can build complex structures, such as starch.

4. ATP molecules are units of energy that can be used for a variety of functions such as movement, transport, or forming chemical bonds.

5. A metabolic pathway is a series of reactions in which the product of one reaction becomes the substrate for the next.

6. Coenzymes help to align the substrate with the active site of an enzyme. Many vitamins act as coenzymes in humans.

Multiple Choice

1. e	8. b	15. c	22. b
2. c	9. b	16. d	23. d
3. b	10. b	17. e	24. c
4. d	11. d	18. a	25. a
5. c	12. a	19. b	
6. c	13. b	20. b	
7. b	14. a	21. c	

What's It All About?

1. **Type of question:** Already defined.

2. **Collect the evidence:**
   a. First law of thermodynamics says that energy is not created or destroyed, only transformed. Transformation is the key point. Provide an example of how cells transform energy, such as the glycogen example provided in the chapter.
   b. Second law of thermodynamics says that energy transfer results in increased disorder in the universe, which is not the same thing as a system within that universe. So a system can harvest some of the energy in the universe, but not all. The energy that is dissipated into the universe increases entropy (key term), so the second law is obeyed.

3. **Pull it all together:**
   According to the first law of thermodynamics, energy cannot be created or destroyed, only transformed. When we talk about cells making energy in the form of ATP, what we really mean is that they transform some other form of energy, such as the chemical energy in the bonds holding glucose molecules together as glycogen, into the chemical energy of phosphate bonds in ATP. During this transformation, some of the energy is transformed into heat. At the end of the transformation, the same amount of energy exists; whereas all of the energy was formerly in one type of chemical bond, however, now it exists as another type of chemical bond plus heat.

   The second law of thermodynamics states that the universe tends to become more disordered during these energy transformations, a state known as an increase in entropy. Even though cells are highly ordered systems, the making of these systems has converted solar energy (the energy of the universe) into chemical energy and heat. Dissipating this energy as heat means that we cannot go backward from chemical energy to sunlight energy, so the total entropy of the universe has increased. Therefore, even though living organisms appear to violate the laws of thermodynamics, when these energy transformations are considered on the scale of the universe, we can show that living things do obey the law.

CHAPTER 7 Vital Harvest: Deriving Energy from Food

Compare and Contrast

oxidation/reduction

Both are reactions involving the transfer of electrons. However, in oxidation electrons are lost, whereas in reduction electrons are gained.

mitochondria/cytoplasm

Both are sites of energy harvesting. However, glycolysis takes place in the cytoplasm, whereas the Krebs cycle and the ETC take place in the mitochondria.

pyruvic acid/citric acid

Both are molecules produced during cellular respiration. However, pyruvic acid is produced by glycolysis, whereas citric acid is part of the Krebs cycle.

cellular respiration/glycolysis

Both are cell processes that result in energy production. However, cellular respiration occurs only in eukaryotic cells, whereas glycolysis can occur in both eukaryotic and prokaryotic cells.

electron transport chain/NAD

Both the electron transport chain and NAD are involved in the last stage of glycolysis, which yields the greatest amount of energy. However, the electron transport chain is a process, whereas NAD is a molecule that has the unique capability of transporting electrons "down" the electron transport chain.

Short Answer

1. ATP is necessary to start the first stage of glycolysis. When ATP donates a phosphate group to glucose, the whole process begins. Eventually, the ATP molecules that are "donated" are replenished.

2. *Coupled reactions* is a term used to describe situations in which the product of one reaction becomes the substrate of the other. Redox (reduction/oxidation) reactions are prime examples of this.

3. To be very efficient at producing energy, an artificial cell would have to have a component similar to a mitochondrion. It would have to be able to harvest the energy from electrons in a gradual process so that excess heat energy is not generated. Another necessary component of the artificial cell would be a series of fine membranes (similar to the cristae of the mitochondrion) that would be able to increase the surface area of such a very small piece of machinery.

4. Cells that utilize glycolysis as their only source of energy production are prokaryotic cells (bacteria) or unicellular organisms (such as yeast). These cells are very small and thus have a rather small energy budget.

5. Evolution of the Krebs cycle allowed organisms to harvest energy more efficiently. This in turn would have provided the resources needed to build larger, more complex structures.

6. The movement of electrons through the electron transport chain is what provides most of the ATP that is produced through aerobic respiration. As the electrons are moved through the carriers, hydrogen ions (protons) are also pumped across the membrane. It is actually the hydrogen ions that drive the synthesis of ATP molecules.

7. The Atkins diet is based on low-carbohydrate and high-protein foods. Therefore, your body switches from burning carbohydrates to burning fats and proteins. See Figure 7.10.

Multiple Choice

1. c	8. c	15. a	22. b
2. c	9. e	16. b	23. d
3. c	10. a	17. a	24. a
4. d	11. b	18. b	25. a
5. d	12. c	19. e	
6. b	13. c	20. c	
7. d	14. a	21. f	

What's It All About?

1. **Type of question:** Already defined.

2. **Collect the evidence:**
   a. What biomolecules have been consumed?

Turkey—mostly protein, some fat

Vegetables—mostly complex carbohydrates; could have some fat, depending on preparation.

Dressing, pie—mostly simple carbohydrates and fat.

b. What is the fate of these molecules?
Carbohydrates—complex carbohydrates will be broken down to simpler molecules, which will enter cellular respiration and be broken down to CO_2, water, and the energy molecules, NADH and ATP.

Fats—these are also broken down to glycerol and fatty acids and are oxidized in cellular respiration.

Proteins—probably broken down to amino acids and used to make new proteins.

3. Pull it all together:
Our Thanksgiving feast is made up of various biomolecules: protein and fat in the turkey, mostly carbohydrates in the dressing and vegetables, and mostly fats and simple sugars in the pie. The fats and carbohydrates can be broken down to simpler sugars and oxidized in cellular respiration to make ATP energy. The proteins will probably be broken down to amino acids, which will be used to make new proteins. Because we are usually not very active following Thanksgiving dinner, and because it tends to be a large meal, we will probably use this energy to build new molecules, such as triglycerides or proteins, rather than using the ATP energy to power muscle activity.

CHAPTER 8 The Green World's Gift: Photosynthesis

Compare and Contrast

chloroplast/chlorophyll

Both are structures involved in photosynthesis. However, chlorophyll is a molecule, and chloroplasts are organelles.

photorespiration/photosynthesis

Both processes are related to the conversion of the sun's energy to chemical bonds. However, photosynthesis produces carbohydrates, whereas photorespiration does not.

stroma/grana

Both are parts of the chloroplast. However, grana are stacks of thylakoids, whereas the stroma is the fluid surrounding them.

reaction center/rubisco

Both are involved in photosynthesis. However, the reaction center is required during the light-dependent reactions, whereas rubisco is used during the light-independent reactions (Calvin cycle).

Calvin cycle/CAM photosynthesis

Both are adaptations to warm climates. However, CAM plants shift some activities to night, whereas C_4 plants do not.

Short Answer

1. Because energy can be only transferred and not created, biological systems need a constant influx of energy. That energy comes from the sun and is fixed into organic molecules via photosynthesis.

2. Photosynthesis is driven by the transfer of excited electrons from one molecule to another. These electrons are excited by UV radiation from the sun.

3. The reaction center, made up of chlorophyll molecules, transforms light energy into chemical energy. This occurs during the light-dependent reactions.

4. Splitting of water to replace excited electrons in the reaction center causes the release of hydrogen ions (H^+) and molecular oxygen (O_2).

5. RuBP is the first step in the Calvin cycle, and it is responsible for carbon fixation.

6. Photorespiration occurs on hot, dry days when the enzyme rubisco binds oxygen instead of carbon dioxide, thereby preventing the formation of sugar during the Calvin cycle.

7. CAM overcomes photorespiration by changing the timing of carbon dioxide intake to evening, when photorespiration is less likely to occur. C_4 metabolism moves the Calvin cycle into the middle of the leaf to decrease rubisco exposure to oxygen.

Multiple Choice

1. b	8. d	15. d	22. d
2. d	9. b	16. c	23. e
3. d	10. c	17. c	24. e
4. e	11. b	18. b	25. a
5. d	12. e	19. c	
6. c	13. a	20. c	
7. d	14. b	21. e	

What's It All About?

1. **Type of question:** Already defined.

2. **Collect the evidence:**

Differences	Similarities
Photosynthesis occurs only in plants.	Both use an electron transport.
Photosynthesis makes sugar, but does not break it down.	Both make a proton gradient to make ATP.
Photosynthesis makes a reduced electron carrier, and cellular respiration makes an oxidized one.	Both occur in specialized organelles. Both use membrane proteins.

3. **Pull it all together:**

On the surface, photosynthesis and cellular respiration are opposing pathways. Cellular respiration breaks down sugars to make ATP energy, whereas photosynthesis collects the energy in sunlight to make the ATP energy used to make sugar. Cellular respiration is an oxidative process, oxidizing sugars and producing an oxidized electron carrier. In contrast, photosynthesis is reductive, making NADPH from $NADP^+$ and sugars from CO_2. Despite these differences, both pathways share several common features. Both occur in specialized organelles, the chloroplast for photosynthesis and the mitochondria for cellular respiration. Both pathways depend on integral membrane proteins to create a proton gradient that is used to make ATP. In both cases, the movement of electrons is essential to making the final products. Thus, although the details differ, the pathways of photosynthesis and cellular respiration show the same pattern of organization.

CHAPTER 9   Genetics and Cell Division

Compare and Contrast

chromatin/chromosome

Both are involved in packaging DNA in the cell. However, chromatin is any DNA-protein complex, whereas a chromosome is one piece of DNA plus proteins.

microtubule/mitotic spindle

Both are active in cell division. However, microtubules are protein fibers that move the chromosomes around, whereas the mitotic spindle is the football-shaped cage that the microtubules form.

cell cycle/interphase

Both the cell cycle and interphase refer to the repeating pattern of growth, genetic duplication, and division in cells. However, the cell cycle is the general name for the whole process, whereas interphase is the part of the cell cycle exclusive of the M (mitotic) phase.

chromatid/homologous chromosomes

Both terms refer to the chromosomes of an organism. However, a chromatid is one of the two identical strands that make up a duplicated chromosome, whereas homologous chromosomes are the maternal and paternal copies of a specific chromosome.

binary fission/mitosis

Both are types of cell division. However, binary fission occurs in bacteria, and mitosis occurs in eukaryotic cells.

Short Answer

1. DNA is a molecule that is similar to a recipe with specific ingredients and directions. The recipe, if followed exactly, results in a luscious cake. In the human body, DNA is found in the nucleus of all cells, and its chemical makeup specifies the directions that result in a human being.

2. Cells don't simply get larger, because as a cell increases in volume, the surface area does not increase at the same rate. Eventually, the volume will be such that the cell cannot function efficiently to move substances in and out. Therefore, it is a wise investment of energy for a cell to divide to produce more cells that are smaller and more efficient.

3. A large amount of DNA can fit into a single cell because DNA wraps around histones, which are special proteins that can bind DNA. Each histone complex (there are five of them) can bind about 200 base pairs of DNA.

4. Prophase: Chromosomes condense; mitotic spindle starts to form; nuclear envelope breaks down.
   Metaphase: Chromosomes line up on the metaphase plate.
   Anaphase: Sister chromatids separate and move to opposite sides of the cell.
   Telophase: Chromosomes start to decondense; nuclear envelope starts to re-form.
   Cytokinesis: Cell divides into two equivalent daughter cells.

5. Plant cells have rigid cell walls and can't be "belted" like animal cells.

See Figure 9.12 in your textbook for diagram. Cell plate is unique to plants; animal cells form a cleavage furrow.

Multiple Choice

1. c	8. b	15. d	22. b
2. c	9. e	16. d	23. b
3. e	10. c	17. d	24. e
4. a	11. d	18. d	25. b
5. d	12. a	19. a	
6. b	13. b	20. b	
7. d	14. e	21. a	

What's It All About?

1. **Type of question:** Compare and contrast.

2. **Collect the evidence:**
   From Chapter 3, we know that proteins are very large and made up of 20 different kinds of monomeric units (amino acids). Depending on how the amino acids are arranged in the primary sequence, we can imagine an almost infinite number of different kinds of molecules. From this chapter, we know that DNA is made up of only four kinds of monomers. However, it can replicate itself using each strand as a template.

3. **Pull it all together:**
   Proteins are assembled from 20 possible amino acids. If each amino acid represented a unit of information, then the primary sequence of a protein could carry information, much like words in a sentence carry information. As

information carriers, protein sequences have the advantage of almost limitless variability. The disadvantage lies in how to copy the sequence exactly without other proteins. Although much simpler in structure because it has only four possible nucleotides, DNA has a built-in mechanism for making exact copies of itself. This self-replicating feature of DNA gives it a great advantage over proteins as an information-storage molecule.

CHAPTER 10 Preparing for Sexual Reproduction: Meiosis

Compare and Contrast

chromosome/chromatid

Both are linear DNA sequences with associated proteins. However, a chromosome consists of one chromatid before it is replicated; after replication, there are two chromatids.

haploid/diploid

Both refer to the number of chromosomes present in a cell. However, haploid means having one set of chromosomes (the number found in gametes), whereas diploid refers to having two haploid sets of chromosomes (the number found in somatic cells).

meiosis/mitosis

Both are types of cell division. However, meiosis reduces the chromosome number and generates diversity, whereas mitosis generates identical daughter cells.

spermatocytes/oogonia

Both are cells produced during gametogenesis. However, spermatocytes are involved in male gametogenesis, whereas oogonia are involved in female gametogenesis.

meiosis I/meiosis II

Both processes are required to produce gametes. However, crossing over occurs only during meiosis I.

Short Answer

1. Meiosis is much less common. Mitosis is required for the day-to-day maintenance of the body, whereas meiosis is used only for reproduction.

2. Meiosis II is required to separate sister chromatids and create haploid gametes.

3. Meiosis is necessary for sexual reproduction. Without meiosis, offspring would be polyploid, having multiple sets of each chromosome.

4. The presence (or absence) of the Y chromosome determines the sex of the fetus.

5. Sperm production is continuous and does not require the formation of a functional cell.

6. Asexual reproduction decreases the cost of finding a mate.

Multiple Choice

1. c	8. e	15. d	22. a
2. d	9. a	16. b	23. b
3. d	10. c	17. a	24. a
4. a	11. a	18. c	25. a
5. b	12. d	19. d	
6. c	13. a	20. b	
7. c	14. e	21. b	

What's It All About?

1. **Type of question:** Describe the effect of "A" on "B." In this case, "A" is the process of meiosis, and "B" is chromosomal missegregation.

2. **Collect the evidence:**

The first step is to review gamete formation in humans. If you look at Figure 10.5, you should be struck by how similar the processes look in the male and female; the most significant difference is that in the female, meiosis I and II produce cells that are not all equal in size. But should size matter? Isn't the segregation process still the same? Segregation is the same, but what about the source of these cells and the timing? Remember that missegregation happens more frequently in older humans.

3. **Pull it all together:**

Gamete formation in human males and females differs in two significant ways. First, gamete formation in males produces four spermatids of equal size, but in females only a single egg is produced; the other three polar body cells are degraded. If missegregation happens in either meiosis I or II, it will affect only one sperm among millions but in females could occur in the single cell that will become the oocyte. Second, the nature of the starting cells is very different. In males, sperm are produced from spermatogonia, which have the capacity to regenerate themselves. Because of this self-renewing property, males are capable of making sperm throughout their lives. The implication is that sperm are always "made fresh," meaning that each spermatogonium hasn't been waiting around and aging. In contrast, female gametes originate from oogonia, which are not self-renewing. Every oogonium in a female ovary has been there since embryonic life stuck in meiosis I, meaning that by the time meiosis is completed and an egg is produced, the egg is at least 12 to 13 years old. Because missegregation happens more often as people age, it seems likely that the female's eggs are more likely to be making errors in meiosis II.

CHAPTER 11 The First Geneticist: Mendel and His Discoveries

Compare and Contrast

genotype/phenotype

Both are genetic terms to describe facets of an organism. However, genotype refers to genetic composition, whereas phenotype refers to the physical appearance.

allele/gene

Both describe a DNA sequence that encodes instructions. However, the actual instructions are found in the gene, whereas the different variants of a gene are referred to as alleles of that gene.

Law of Segregation/Law of Independent Assortment

Both refer to the behavior of genes and chromosomes during meiosis. However, the Law of Segregation states that different alleles of the same gene move away from one another during meiosis, such that each gamete gets only one of the two alleles, whereas the Law of Independent Assortment states that gene pairs assort independently from one another.

dominant/recessive

Both describe the behavior of alleles. However, dominant alleles are always observed in the phenotype, whereas the phenotype of a recessive allele is observed only if two recessive alleles (and no dominant alleles) are present.

rule of addition/rule of multiplication

Both terms refer to probability rules that are used to predict genetic outcomes. However, the rule of addition is used when there is an outcome that can occur in two or more ways, whereas the rule of multiplication is used to predict the outcome of two or more things occurring simultaneously.

Short Answer

1. Mendel was a mathematics teacher by training and he was a very thorough scientist. He kept meticulous notebooks, replicated his studies many times and then used statistical analysis to document the ratios of the courses. Prior to Mendel, there were no long-term studies that could be replicated.

2. The genotype is the mutation in the chloride channel gene; the phenotype is the cystic fibrosis disease (and accompanying mucus production in the lungs).

3. Orange is dominant. The ratio in F_2 would be 3 orange: 1 yellow.

4. a.

	C	C
C	CC	Cc
c	Cc	cc

    b. On average, one child in four (1/4 chance for each child) would be expected to be homozygous for the cystic fibrosis channel-gene mutation (*cc*).

5. Pleiotropy, in which one allele affects multiple systems in the body.

6. Although she may have the mutation, she may never develop breast cancer because the many activities and biochemical functions of life are tied together in a complex interrelationship.

7. Peas and fruit flies are good subjects for genetic investigations because they are easy to grow or culture, they have rapid breeding times, and they have easily visible traits.

8. If two pink snapdragons were crossed, the ratio would be 1 red : 2 pink : 1 white.

9. In the first case, the rule of multiplication holds in that probability of any two events happening is the product of their probabilities: $\frac{1}{2} \times \frac{1}{2} = \frac{1}{4}$. In the second case, the rule of addition applies, so that the probability is the sum of the individual probabilities: $\frac{1}{4} + \frac{1}{4} = \frac{1}{2}$.

10. Factors others than genetics influence IQ, such as lifestyle and the social and intellectual environment that a child is exposed to while growing up.

11. One might ask the following: (a) What is the usual pattern of inheritance? (b) Should my other family members be tested? (c) Does this mean my children will get the disease?

Multiple Choice

1. c	8. b	15. d	22. c
2. e	9. c	16. c	23. c
3. a	10. d	17. a	24. d
4. b	11. b	18. b	25. c
5. d	12. c	19. b	
6. b	13. c	20. a	
7. c	14. c	21. d	

What's It All About?

1. Type of question: Describes the effect of "A" (linked traits) on "B" (Mendel's analysis).

2. Collect the evidence:
- What did the ratios tell Mendel? There is a lack of blending in inheritance, such that traits are defined as dominant and recessive.
- Think about what the F_1, F_2, and F_3 generations would look like for a cross of green wrinkled (*yyss*) with yellow smooth (*YYSS*) if the *Y* and *S* genes never separated, so *YS* and *ys* always go together in the gametes. How would this be different from unlinked traits?

3. Pull it all together:
Mendel's Law of Independent Assortment arose from his observation that the peas never "lost" a trait; that a trait that seemed to disappear in the F_1 generation always reappeared intact in the F_2. Even in the case of linked traits, Mendel would have seen a similar pattern in the F_1 generation (all yellow and smooth seeds) and the recessive traits reappearing in F_2, although he probably would have been puzzled why he had a ratio of 3 yellow, smooth: 1 green, wrinkled seeds instead of 9:3:3:1 he had observed for the other five traits when examined in paired combinations. This unexpected finding might have caused him to question his Law of Segregation, because he would never see the trait for seed color separate from seed shape. Given how carefully Mendel kept records, and his proven insight into the mechanism of inheritance despite knowing nothing about genes, I believe he would have concluded that seed color and shape were a single trait and not two.

CHAPTER 12 Units of Heredity: Chromosomes and Inheritance

Compare and Contrast

aneuploidy/polyploidy

Both refer to abnormal chromosome numbers. However, aneuploidy is an abnormality of one or a few chromosomes, whereas polyploidy refers to having an extra haploid set of chromosomes.

inversion/translocation

Both are structural abnormalities of chromosomes. However, an inversion is a flipping of a sequence within a chromosome, whereas a translocation is an exchange between non-homologous chromosomes.

Turner/Klinefelter

Both are sex-chromosome abnormalities. However, Turner syndrome patients have only one X chromosome, whereas individuals with Klinefelter syndrome have two X chromosomes and a Y chromosome.

sex-linked recessive/autosomal recessive

Both are inheritance patterns for genetic disorders. However, in sex-linked recessive inheritance, the altered gene is on the X chromosome, whereas in autosomal recessive inheritance, the altered gene is an autosome (chromosomes 1 through 22 in humans).

autosome/sex chromosome

Both are types of chromosomes. However, the autosomes can be found in either sex, whereas the sex chromosomes are sex-specific (XX or XY).

Short Answer

1. Because male humans have only one X chromosome, they are more likely to express traits that are recessive in females who have two X chromosomes.

2. First, the disease would become much more common in the population. Then, depending on its age of onset, it could end up killing all the heterozygotes as well as homozygotes, thereby disappearing altogether from the population.

3. Refer to Figure 12.6.

4. Polyploidy is very rare; in fact, in humans there are no known cases of individuals surviving much beyond birth as polyploid.

5. Cancer occurs as a result of unchecked mitosis. Some research studies have demonstrated that certain kinds of cancer cells are more likely to have nondisjunctions than are normal cells.

6. A translocation is the movement of genetic material from one chromosome to another, whereas nondisjunction is a failure of chromosomes to sort properly during meiosis, resulting in aneuploidy (an unusual chromosome number).

Multiple Choice

1. a	8. a	15. a	22. a
2. a	9. d	16. d	23. e
3. c	10. d	17. c	24. e
4. c	11. c	18. d	25. d
5. d	12. b	19. c	
6. e	13. e	20. c	
7. e	14. b	21. d	

What's It All About?

1. **Type of question:** Defend a position—in this case, the position has been defined by the observation that mature twins are not indistinguishable.

2. **Collect the evidence:**

Think about what you've learned in this chapter about inheritance to make some predictions about the genotype of these twins. Then think about the definitions of *genotype* and *phenotype,* which you learned in Chapter 11. Is genotype the only factor that can affect phenotype?

3. **Pull it all together:**

Because identical twins result from a single egg giving rise to two fetuses, this is a case of natural cloning. Both children will have identical genotypes, which we could predict to have identical phenotypes. However, identical twins provide compelling evidence that the environment ("nurture") may have significant effects on appearance and behavior ("nature"). Just as plants exposed to different growth conditions of altitude, water, and food can grow to be different sizes, so too can differences in the environment affect the development of people. Even twins growing up in the same household cannot have exactly the same number and kind of experiences; at some point they will have unique experiences, such as being placed in different classrooms in school, which may affect future behavior. Because of their shared genes, they will probably always look alike, share common mannerisms, and have similar preferences for activities, but differences in how they are treated by family, teachers, and friends will shape their personalities differently.

CHAPTER 13 Passing on Life's Information: DNA Structure and Replication

Compare and Contrast

germ-line cell/somatic cell

Both are cell types found in the human body. However, a mutation in a germ-line cell will be passed on to all generations, whereas a mutation in a somatic cell will not be passed on to future generations.

mutation/point mutation

Both are types of mutations. However, the term mutation refers to the permanent alteration of DNA base sequence, whereas a point mutation is a specific type of mutation in which a single base pair is altered.

purine/pyrimidine

Both are types of nitrogen bases found in DNA. However, a purine base (adenine, guanine) is a double-ring molecule, whereas a pyrimidine base (thymine, cytosine) is a single-ring molecule.

nucleotide/DNA

Both are components of nucleic acids. However, a nucleotide consists of a phosphate, sugar, and nitrogen base, whereas DNA is a specific type of nucleic acid consisting of repeating units of nucleotides.

DNA polymerase/DNA replication

Both are essential components to synthesizing DNA. However, DNA replication is the process, whereas DNA polymerase is the enzyme essential to the process.

Short Answer

1. Prior to the 1930s, biologists were concerned about the chemical nature of genes and exactly what they did in cells. The investigations of Beadle and Tatum with the bread mold *Neurospora* indicated that mutations in genes were linked to alterations in enzymes produced. The investigation was eventually labeled as the "one-gene, one-enzyme" hypothesis.

2. Rosalind Franklin and Maurice Wilkins, who provided the X-ray diffraction patterns.

3. Replication is not perfectly accurate. However, the editing activity of DNA polymerase helps to correct mistakes.

4. No; because the gametes did not have the mutation, only the skin cell and its daughter cells will have the mutation. It cannot be passed on through the germ line.

Multiple Choice

1. b	8. b	15. c	22. c
2. b	9. d	16. c	23. b
3. a	10. c	17. b	24. a
4. d	11. b	18. c	25. a
5. c	12. a	19. c	
6. b	13. e	20. b	
7. c	14. b	21. a	

What's It All About?

1. **Type of question:** Compare and contrast—how are mutations in gametes different from those that arise in somatic cells?

2. **Collect the evidence:**
   From this chapter, we know that one mutation is usually not enough to produce a phenotype such as cancer; rather, several mutations are needed. Also, from Chapters 11 and 12, we've learned about pleiotropy and the types of genetic aberrations that produce disease.

   DNA replication—extremely accurate, so spontaneous mutations are very rare.

   Mitosis—also extremely accurate, so a rare mutation will be propagated faithfully in somatic cells.

   Gametogenesis—is accomplished through meiosis, so it is a reductive process, meaning that if a mutation happens during DNA replication in meiosis I, then only one of the four gametes produced in meiosis II will carry the mutation.

   Mutations—can happen spontaneously during DNA replication, but can also be induced by outside agents, such as UV light, cigarette smoke, toxins, or viruses.

3. **Pull it all together:**
   Changes in phenotype caused by changes in the DNA generally require more than a single mutation event, which decreases the odds that we ever see the result of a mutation. Heritable mutations are those that happen during the process of meiosis that creates the gametes. Presuming that a mutation happens during meiosis I and is not repaired, only one of the four gametes produced would carry the mutant gene. Assuming the mutation is recessive, this mutant gamete would have to fuse with another gamete carrying a mutation in the same gene for a defective phenotype to be expressed in the organism. When one considers the huge number of gametes that organisms produce, the rarity of spontaneous mutations, and the low probability that two mutations in the same gene would end up in the same fertilized egg, one can understand why heritable mutations are so rare.

   Mutations in somatic cells, although still rare, are passed along to each daughter cell during mitosis. During subsequent divisions, the mutation will be preserved in the DNA, so there is no way to "lose" the mutation. Also, if these mutant cells accumulate other mutations over time, then the odds that we will see a change in phenotype will increase. So even though spontaneous mutations are rare, the accuracy of DNA replication in mitosis and life span of somatic cells means that it is more likely to develop cancer in a somatic cell rather than a new heritable mutation.

CHAPTER 14 How Proteins Are Made: Genetic Transcription, Translation, and Regulation

Compare and Contrast

transcription/translation

Both are involved in expressing a gene. However, transcription converts the information from DNA to mRNA, whereas translation produces the protein product.

tRNA/mRNA

Both are types of RNA produced by transcription. However, tRNA is involved in bringing amino acids to the growing polypeptide, and mRNA encodes the order of amino acids.

operator/repressor

Both are involved in operon regulation in bacteria. However, the operator is a DNA sequence to which the protein repressor binds.

intron/exon

Both are parts of eukaryotic genes. However, introns do not code for protein, whereas exons do.

codon/genetic code

Both are terms used to describe the genome. However, a codon is the genetic code for just one amino acid.

Short Answer

1. Through their work with bacteria, Jacob and Monod were the first to demonstrate a mechanism by which protein synthesis (that is, gene expression) could be controlled.

2. Proteins can be quite long, which means that there are many, many different possible combinations of amino acids.

3. The anticodon is a sequence located on the transfer RNA that matches the amino acid-defining codon located on the mRNA. The anticodon-codon binding allows the proper matching of amino acid and nucleotide sequence. A mutation in this sequence could result in a failure of tRNA to bind the appropriate mRNA codon leading to a change in protein sequence.

4.

Molecule	Location	Structure	Function
DNA storage	nucleus	deoxyribose, phosphate, ATGC	information (gene)
RNA	nucleus and cytoplasm	ribose, phosphate, storage AUGC	messenger, transfer, ribosomal functions

5. MicroRNAs are regualtory structures that interfere with the translation of RNA into protein. They hold the potential not only to explain current observation but also to provide new tools for biotechnology.

6. Without our diverse genome and the mechanisms used to interpret it, life as we know it would not be possible.

Multiple Choice

1. b	8. d	15. a	22. d
2. c	9. b	16. d	23. b
3. d	10. a	17. c	24. a
4. e	11. b	18. c	25. a
5. a	12. c	19. e	
6. e	13. b	20. c	
7. c	14. d	21. c	

What's It All About?

1. **Type of question:** Variation on "defend a position"—asks you to explain an observation.

2. **Collect the evidence:** Think DNA structure, particularly base composition.

3. **Pull it all together:**
   The sequence of nucleotides in the mRNA that an organism needs to translate into protein depends on the sequence of nucleotides in the DNA. It is possible that different organisms might have a bias in their DNA so that there are more AT nucleotide pairs than GC pairs. More AT pairs might increase the frequency of uracil in the mRNA, so there would be slightly more triplet pairs that include a U, biasing the code.

CHAPTER 15 The Future Isn't What It Used to Be: Biotechnology

Compare and Contrast

reproductive cloning/therapeutic cloning

Both are techniques to make a copy of something. However, in reproductive cloning, an entire organism is exactly duplicated; in therapeutic cloning, embryonic stem cells are used to treat disease.

cloning vector/plasmid

Both are involved in cloning (copying) a gene. However, a vector is a type of molecule, which carries the gene into host cells, and a plasmid is a particular type of cloning vector.

blastocyst/embryonic stem cells

Both refer to cells critical to genetic engineering. However, a blastocyst is a stage in embryonic development from which embryonic stem cells can be harvested. The embryonic stem cells can then be placed in another organism.

recombinant DNA/transformation

Both are processes that involve transfer of DNA. However, recombinant DNA is the final product, whereas transformation is the process by which a bacterium incorporates DNA from outside its cell, resulting in recombinant DNA.

DNA profiling/short tandem repeats (STR)

Both terms refer to the use of DNA in forensic biotechnology. However, DNA profiling is the process of matching up DNA from different persons, whereas short tandem repeats refer to DNA patterns that vary from person to person. This matching of short tandem repeats is crucial to DNA profiling.

Short Answer

1. The universal genetic code allows human proteins to be made in bacteria (see Chapter 14).

2. Restriction enzymes cut both strands of a DNA molecule at specific sites.

3. Cloning is making an exact copy. To clone your best friend or dog:
   a. Remove a cell from your friend.
   b. Fuse that cell with an egg.
   c. Let the embryo develop.
   d. Implant the embryo in a surrogate mother.
   e. Wait for your clone to be born.

4. PCR has allowed the analysis of minute quantities of evidence and has facilitated DNA fingerprinting.

5. Consider the sources and availabilities of these cells. This is a good question to discuss with your friends and classmates.

6. People distrust genetically modified organisms because they do not understand how they were produced and because there are conflicting messages regarding their safety. The moral and ethical issues surrounding such areas as xenotransplantation, human cloning, and genetically modified food require that time and money be spent to understand the impact of these techniques on our society.

Multiple Choice

1. c	8. c	15. b	22. b
2. c	9. b	16. c	23. c
3. c	10. e	17. b	24. e
4. d	11. b	18. c	25. c
5. a	12. d	19. a	
6. d	13. c	20. e	
7. a	14. c	21. d	

What's It All About?

1. **Type of question:** Variation on effect of "A" on "B," where "A" is forensic analysis and "B" is the problem of identifying victims of a mass disaster.

2. **Collect the evidence:**
   As a forensic technician, you will use molecular identification methods rather than physical ones, especially because there isn't much useful physical evidence left. What techniques would you use? Why? How are you going to be sure that you are correct in your identifications?

3. **Pull it all together:**
   The bombing of the World Trade Center left few of its victims intact enough for traditional methods of identification (fingerprints, dental records) to be useful; the method of choice would be forensic DNA typing. The first task would be to collect blood from the parents, siblings, or children who lost a loved one in the disaster, because DNA typing requires comparison of DNA patterns from a sample of unknown identity to the patterns in identified samples. From each sample of human remains collected on the site, a sample of DNA would have to be amplified using PCR and cut using restriction enzymes. Fragments would have to be separated by gel electrophoresis and then compared to the patterns generated by the DNA of family members. Every sample taken from the site must be compared against every sample taken from every family who believes they lost someone; thousands of unknowns must be compared against thousands of knowns at multiple DNA sequences to confirm the final identification.

CHAPTER 16　An Introduction to Evolution: Charles Darwin, Evolutionary Thought, and the Evidence of Evolution

Compare and Contrast

paleontology/taxonomy

Both are disciplines that provide evidence for evolution. However, paleontology is the study of (often) extinct animals in the fossil record, and taxonomy is the study of living animals and their distribution.

natural selection/fitness

Both are related to the mechanism of evolution. Natural selection is the mechanism by which evolution occurs, and it depends on the fitness of individual organisms (how well an organism interacts with, and survives in, its environment).

embryology/homologous structures

Both provide evidence for evolution. However, embryology is the study of how animals develop, and homologous structures are similar because of inheritance from a common ancestor.

evolution/natural selection

Both describe how organisms change over time into new species. However, evolution is the process of this change, whereas natural selection is the mechanism by which the change occurs.

Darwin/Wallace

Both had their theories on evolution presented at the Linnaean Society in 1858. However, Darwin went on to publish several books on the subject.

Short Answer

1. *Descent with modification* is the term used to describe the change that occurs in populations as they are sculpted by natural selection from one generation to the next.

2. Thomas R. Malthus wrote an essay that provided Darwin with the basis of his idea on the "struggle for existence."

3. Jean-Baptiste de Lamarck preceded Darwin and theorized on the possibility of descent with modification.

4. On the voyages of the HMS *Beagle.*

5. Alfred Russel Wallace was a naturalist whose travels in Asia led him to the same conclusions as Darwin about natural selection and evolution.

6. *On the Origin of Species by Means of Natural Selection*

7. The modern synthesis incorporated Mendel's work on inheritance to explain the mechanisms by which natural selection could occur.

8. A scientific theory is a logical explanation that fits the available facts about a given phenomenon.

9. Radiometric dating demonstrates the relative age of fossils; fossils show mineralized remains from extinct species; comparative morphology and embryology compare species and reveal their likely relatedness; biogeography studies the distribution of species across geographic regions; gene modification shows the evidence of mutation over time with experimental evidence. Recent experiments with living species demonstrate evolution in action evidence.

Multiple Choice

1. c	8. c	15. e	22. a
2. a	9. a	16. e	23. c
3. c	10. d	17. e	24. e
4. c	11. a	18. b	25. d
5. d	12. c	19. c	
6. e	13. d	20. d	
7. d	14. a	21. e	

What's It All About?

1. Type of question: Variation on "defend a position"—asks you to apply your knowledge to defend a prediction.

2. Collect the evidence:
Genetics—think about allelic variation, meaning that any given gene has enough allelic variants such that phenotypes describe a range of values (such as human height) and not just discrete points.

Natural selection—an individual's genotype allows it to survive and produce offspring under a range of environmental conditions, but any given organism may produce more offspring than other individuals in the population if it is more successful at growing and breeding under the existing environmental conditions.

Common descent with modification—offspring can inherit only those genes present in the parents, but natural selection favors the survival of some individuals over others. Over time, the favored traits predominate until eventually the offspring may look very different from the ancestral organism.

3. Pull it all together:
Within any population of organisms exists variation in the genotypes that produce a range of possible phenotypes. It is this range of phenotypes that forces in the environment act on, with the most favorable forms selected for survival and increased reproduction. As long as the environment remains stable, these phenotypes will be fixed in the population. Global warming is an example of a change in environmental conditions that would drive selection of more heat-tolerant organisms. I would predict that if the regions of tropical climate on Earth increased, then those animals among the current populations that could best adjust to this change would survive. For example, rabbits with thin fur or tough skins might be more successful in a warmer climate; these phenotypes would eventually replace the current phenotypes we know for rabbits.

CHAPTER 17 The Means of Evolution: Microevolution

Compare and Contrast

microevolution/macroevolution

Both are evolutionary processes. However, microevolution is a change in the allele frequency in a population, whereas macroevolution is a dramatic change in a population eventually leading to a new species.

gene flow/genetic drift

Both are agents of microevolution. However, gene flow is the movement of alleles between populations, whereas genetic drift is the chance change in allele frequency (e.g., through the reduction in population size in a bottleneck effect).

gene pool/population

Both refer to groups of individuals. However, the gene pool is the totality of alleles present in a population, whereas a population is a group of individuals living together and reproducing.

fitness/adaptation

Both refer to how well organisms interact with their environments. However, fitness refers to the number of offspring individuals leave, whereas adaptation is the change in organisms over time to better fit their environment.

population/species

Both population and species are terms important to population genetics. However, a population consists of all members of a species that live in a specific geographic region, whereas a species refers to those members of a population that are capable of interbreeding.

Short Answer

1. The driving force behind microevolution is mutation, because it changes the frequency of alleles in a population.

2. The individual with the greater reproductive success (larger number of surviving offspring) is more fit.

3. Sexual selection is a form of nonrandom mating. If mating is not random, then individuals will have different levels of reproductive success, which will in turn change gene-pool frequencies.

4. Inbreeding is not an evolutionary mechanism because it reduces allelic frequency and literally simulates the bottleneck effect. No new variations are brought into the population.

5. The answer to this problem varies: (1) the field mice could begin to adapt to the presence of humans and even cohabit; (2) the increased numbers of domestic pets might force the population to migrate; (3) the population might split, with some staying in the area and others looking for new territory. You may even develop some other scenarios.

Multiple Choice

1. e	8. c	15. e	22. a
2. b	9. c	16. d	23. d
3. a	10. c	17. a	24. c
4. c	11. a	18. a	25. c
5. d	12. b	19. b	
6. d	13. a	20. b	
7. c	14. c	21. c	

What's It All About?

1. **Type of question:** Variation on "defend a position"—asks you to apply your knowledge to defend a prediction.

2. **Collect the evidence:**
   Mutations—mutations can be major events, such as when chromosomal structure changes (Chapter 12—inversions, deletions, rearrangements), or subtler, such as point mutations. In either case, they are generally neutral or bad. But if a mutation occurs in a gene that is pleiotropic (Chapter 11), the mutation could affect many genes at the same time.

   Natural selection—mutations are the fuel that drives natural selection (this chapter). A mutation, or collection of them, that changes an organism's phenotype such that the change confers some advantage would allow that organism to survive better and produce more offspring, increasing the frequency of the phenotype.

3. **Sample answer:**
   When one looks at any complex structure in a living organism, it is hard to imagine how accumulating mutations could produce an eye. However, genes code for molecules that execute functions, such as enzymes, as well as for

molecules that control how, when, and where these functions are executed. A point mutation in the gene for protein that binds a light-sensitive pigment might cause the protein to interact with itself so that cluster of protein-bound pigments forms in the membrane, making it sensitive for light. If light sensitivity helps the organism survive, maybe by avoiding predators or finding food more successfully, there would be positive selection pressure to keep this mutation. Additional mutations, maybe in genes that control where the pigment-binding molecules are expressed, or the shape of the structure holding these molecules, could lead to a structure that detects the direction of the light. These modifications would be retained if they proved helpful. Thus, it is possible that mutations accumulating over time could eventually produce an eye.

Conversely, a specialized structure could develop suddenly if a large number of mutations occurred in the organism almost simultaneously, say as the result of mutagen exposure. Although statistically possible, the likelihood that a large number of random events happened almost simultaneously with such a favorable outcome seems very low.

CHAPTER 18 The Outcomes of Evolution: Macroevolution

Compare and Contrast

cladogenesis/anagenesis

Both refer to modes of speciation (how species arise). However, in cladogenesis, new species branch off from a continuing species, whereas in anagenesis a new species arises without branching from the ancestral species.

extrinsic isolating mechanism/intrinsic isolating mechanism

Both refer to how different species may be reproductively isolated. However, extrinsic mechanisms are not part of the organisms' makeup (e.g., geographic separation by large rivers), whereas intrinsic mechanisms are part of the organisms' makeup (e.g., behavioral patterns that prevent mating with the wrong species).

phylum/kingdom

Both are levels of biological classification. However, a phylum is a level of classification below that of kingdom.

species/population

Both refer to groups of organisms. However, populations are subgroups of a species.

systematics/cladistics

Both are methods for grouping species. However, cladistics is based on shared derived characteristics to determine evolutionary relationships among species.

Short Answer

1. Sympatric speciation occurs within the same geographic area or shared habitat, whereas allopatric speciation takes place in populations separated by some barrier.

2. Adaptive radiation is the appearance of a series of new species that arise because of the arrival of a single (ancestral) species in a new habitat.

3. The pace of evolutionary change is the major difference between the two models.

4. The number of shared derived characteristics is used to design phylogenetic trees and therefore is the result of cladistic analysis.

5. Homologous structures are derived from those found in a common ancestor—for example, an early mammal ancestor to both the gorilla and the bat. Analogous structures represent a common solution (traveling swiftly on land) derived from different sources.

6. New species arise either from the change in an existing species that is so profound that it becomes distinct from its ancestral species, or from a speciation event that splits any existing species into several new populations.

Multiple Choice

1. b	8. c	15. c	22. c
2. a	9. b	16. c	23. d
3. d	10. a	17. e	24. c
4. d	11. c	18. a	25. c
5. c	12. a	19. e	
6. a	13. d	20. c	
7. c	14. a	21. d	

What's It All About?

1. **Type of question:** Variation on "defend a position"—asks you to apply your knowledge to describe an observation.

2. **Collect the evidence:** What drives evolution? What advantages does complexity provide?

3. **Pull it all together:**
   Evolution is driven by the force of natural selection on variations in organismal phenotypes. A phenotype that allows an organism to produce more offspring will become more common in successive generations as long as the environment remains stable. Complex structures allow for new or improved functions that enable organisms to succeed in different environments, such as occurred when ancient amphibians moved onto land. Complex structures can also allow organisms to adapt to changing environments; terrestrial animals can move to new places, whereas fish are limited to their pond. Complexity is not better or worse than simple structures, but it does seem to allow organisms more options for dealing with the environment.

CHAPTER 19 A Slow Unfolding: The History of Life on Earth

Compare and Contrast

angiosperm/gymnosperm

Both are vascular plants that produce seeds. However, gymnosperms are nonflowering plants, whereas angiosperms are flowering plants.

Archaea/Bacteria

Both are kingdoms of unicellular organisms. However, Archaea include the extremophiles.

epoch/era

Both are periods of time. However, eras are longer than epochs.

domain/phylum

Both are classifications of organisms. However, domains are larger (more inclusive) than phyla.

bryophytes/seedless vascular plants

Both are types of primitive plants. However, bryophytes include the mosses and liverworts, whereas seedless vascular plants include the club mosses and ferns.

Short Answer

1. The Malnourished Earth hypothesis refers to a theory that as oxygen built up in the early atmosphere, it also caused the weathering of sulfur from the land and into the ocean. This amounted to a global toxic spill that deprived algae of nutrients and halted their diversification for about 600 million years.

2. The two adaptations were the development of a waxy outer cuticle that protected plants from desiccation and the development of a seed that both housed and fed the developing embryo.

3. Continental drift prevented the colonization of species from one geographic area to another. This in turn would have promoted the adaptive radiation of species into new niches, leading to new adaptations and new species.

4. The impact of an asteroid on the Earth's surface

5. By photosynthesis; oxygen is used for cellular respiration by consumers and also provides the ozone layer for UV protection.

6. Archaea and Eukarya (in spite of having different cell types)

7. The Cambrian Explosion occurred approximately 544 million years ago and lasted approximately 5 million years. It marked the appearance of many new animal forms; almost all (29/30) modern phyla appeared then.

Multiple Choice

1. d	8. d	15. e	22. e
2. d	9. c	16. b	23. b
3. c	10. d	17. a	24. a
4. e	11. a	18. b	25. d
5. c	12. d	19. a	
6. b	13. c	20. a	
7. a	14. b	21. d	

What's It All About?

1. **Type of question:** Variation on "defend a position"—asks you to apply your knowledge to describe an observation.

2. **Collect the evidence:**
   This chapter—life is self-replicating.

   Chapter 18—for a living organism, a self-replicating molecule that can also direct synthesis of other molecules (RNA) is needed.

   Chapter 5—also need a way to separate self from environment: membranes.

   Chapter 7—need a source of energy and the molecules to extract energy from the environment.

3. **Pull it all together:**
   One of the first things needed for a living organism to evolve is a plasma membrane, to concentrate biologically active molecules and to separate them from the environment. I think this is the most time-consuming process because it is limited by diffusion. We believe that the molecules important to living organisms were rare in the early oceans, so it probably took a very long time to capture enough of them inside of a membrane. Once inside a membrane, chemical reactions are more likely to happen. This means that a molecule capable of replicating itself, another criterion of a living organism, is more likely to do so if it is in an environment rich in its monomers. Because chemical reactions are more likely to happen in such a case, the evolution of pathways of reactions has also been speeded up, so a third feature of living organisms can develop. Once these basic requirements of a living thing are met—self-replication, energy assimilation, and interaction with the environment—then changes to the self-replicating mechanism can create new organisms.

CHAPTER 20 Arriving Late, Traveling Far: The Evolution of Human Beings

Compare and Contrast

Australopithecus/Ardipithecus

Both are terms used to describe species of Hominini. However, *Ardipithecus* fossils are much older than most of those of *Australopithecus*.

fossil/skeleton

Both of these represent body structures. However, fossils may be mineralized (or preserved) skeletal remains.

chimpanzee/human

Both are primates and share a high percentage of their DNA sequences. However, they are different species.

arboreal/terrestrial

Both are ecological characteristics. However, arboreal refers to living in (or using) trees, whereas terrestrial refers to dwelling on the ground.

tooth structure/molecular biology

Both are ways of determining evolutionary relationships. However, the extreme age of fossils makes it difficult to reliably use molecular biology to determine their characteristics; so tooth structure tends to be a preferred means in such cases.

Short Answer

1. The Hominini are the group of species that includes modern humans and their ancestral species.

2. It allowed researchers to expand the apparent range of ancient humans in Africa to the west (Chad).

3. To determine evolutionary relationships among Hominini

4. Lucy is a 3.18 Mya fossil of *Australopithecus afarensis* discovered in Ethiopia.

5. Both describe the evolutionary history of humans. However, whereas the "out of Africa" theory cites Africa as the source of modern humans, the "multiregional" theory suggests that multiple migrations occurred from Africa and that these groups later interbred to produce Neanderthals and humans.

6. If it is indeed a distinct Hominini species, its members would have been active while modern humans were established in Europe.

Multiple Choice

1. a	8. d	15. d	22. e
2. b	9. c	16. b	23. a
3. a	10. a	17. a	24. d
4. e	11. d	18. c	25. a
5. a	12. d	19. c	
6. a	13. c	20. d	
7. c	14. a	21. b	

What's It All About?

1. **Type of question:** Variation on "defend a position"—asks you to apply your knowledge to describe an observation.

2. **Collect the evidence:** Examples from the fossil record, dates.

3. **Pull it all together:**
   Paleoanthropology is the study of early humans. Based on the fossils of *H. neanderthalensis, H. sapiens,* and *H. floresiensis,* it seems likely that the first modern humans migrated out of Africa about 50,000 years ago. They may indeed have had contact with both Neanderthals and Hobbit people.

CHAPTER 21 Viruses, Bacteria, Archaea, and Protists: The Diversity of Life 1

Compare and Contrast

autotroph/heterotroph

Both refer to modes of nutrition. However, autotrophs use the sun's energy to synthesize organic molecules from inorganic molecules, whereas heterotrophs consume organic molecules.

phytoplankton/zooplankton

Both are types of protists. However, phytoplankton are autotrophs, whereas zooplankton eat phytoplankton.

viruses/protists

Both can be pathogenic, that is, disease-causing. However, viruses are acellular and therefore not considered to be living, whereas protists are living organisms.

colonial multicellularity/true multicellularity

Both terms refer to organisms that are composed of multiple cells. However, an organism that is colonial (such as the alga *Volvox*) is a loosely linked collection of cells, whereas an organism with true multicellularity has different cells with different functions.

cilia/flagella

Both are appendages that can be used to propel protists. However, cilia are short, oar-like appendages, whereas flagella are long, whip-like appendages.

Short Answer

1. The principle behind vaccination is to allow the body to develop an initial immune response against killed or attenuated viruses. Examples include vaccinations against measles and polio.

2. Bacteria convert nitrogen to a useful form, and they help in decomposition and sewage treatment.

3. Although their cells structurally resemble the Bacteria, Archaea seem to be more closely related to the Eukarya, and many of their genes resemble eukaryotic genes (although some also resemble bacterial genes).

4. Protists (1) are eukaryotic; (2) reproduce sexually; (3) inhabit moist environments; (4) are mostly microscopic; (5) are animal-like, plant-like, or fungi-like.

5. Extremophiles live in conditions of (for example) high salt, high temperature, high pressure, and high or low pH. Venus is a lot hotter than Earth, so organisms there may resemble the archaeal theromophiles and have enzymes adapted to work at high temperatures.

6. Phytoplankton are photosynthetic aquatic organisms that belong to the Domain Prokarya or Kingdom Protista (algae).

7. The Domain Bacteria is most similar to the Domain Archaea in terms of organismal size and organization, although Archaea may also have some ties to the Eukarya as well.

8. Most antibiotics are designed to be effective against bacteria and therefore do not affect viruses.

9. Protists evolved approximately 2 billion years after the archaea and bacteria.

10. Protists can have several roles: (1) producer—the plant-like protists produce carbohydrates and evolve oxygen and are thus autotrophic; (2) consumer—the animal-like protists must capture other organisms and are thus heterotrophic.

Multiple Choice

1. d	8. c	15. a	22. d
2. e	9. e	16. d	23. b
3. d	10. e	17. e	24. e
4. a	11. c	18. d	25. a
5. d	12. a	19. e	
6. e	13. b	20. d	
7. a	14. c	21. c	

What's It All About?

1. **Type of question:** Variation on "defend a position"—asks you to apply your knowledge to describe an observation.

2. **Collect the evidence:**

Why does any organism persist? As long as organisms successfully reproduce—that is, adapt to changes in the environment and survive—they will persist unless something can drive their population to extinction, such as a natural disaster.

Protists are generally single-celled organisms that can reproduce sexually or asexually. Like bacteria, they can exist in large populations, reproduce quickly, and adapt rapidly to changes in the environment.

3. **Pull it all together:**

Organisms will persist as long as they can successfully reproduce and as long as no natural disaster drives the population to extinction. Natural selection doesn't specifically choose complex organisms over simple ones; it selects successful organisms. The Protista, as a Kingdom, are remarkably diverse organisms. They can exist in large populations in aquatic or terrestrial environments. They can reproduce quickly. They have exploited many strategies for finding food. As a group, protists are adaptable organisms, so they persist.

CHAPTER 22 Fungi and Plants: The Diversity of Life 2

Compare and Contrast

autotroph/heterotroph

Both refer to modes of nutrition. However, autotrophs use the sun's energy to synthesize organic molecules from inorganic molecules (such as CO_2), whereas heterotrophs consume organic molecules.

club fungi/imperfect fungi

Both are categories of fungi. However, the term imperfect fungi (or mitosporic fungi) is used for species that have not (yet) been observed to reproduce sexually.

sporophyte/gametophyte

Both terms refer to the alternating generations of plants. However, the gametophyte produces haploid structures, whereas those of the sporophyte are diploid.

angiosperm/gymnosperm

Both are seed-producing plants. However, angiosperms have flowers, but gymnosperms do not.

lichen/mycorrhizae

Both are examples of symbiotic relationships. However, mycorrhizae involve a plant and a fungus, but lichens are composed of a fungus and either an alga or bacterial colony.

Short Answer

1. Bryophyte sperm must swim through water to fertilize the egg.
2. A lichen is a symbiotic association of fungi with either algae or bacteria.
3. Gymnosperms have both seeds and a vascular system.
4. Plants use nectar to attract animal pollinators.
5. Mushroom caps (fruiting bodies) disperse spores.
6. The production of fruit allows plants to use animals as seed dispersers, which thereby benefits both species.
7. Fungal spores can be produced by mitosis (asexual reproduction) or through fusion of nuclei from separate fungi (sexual reproduction).
8. Endosperms, such as rice and wheat, are common food sources for humans.

Multiple Choice

1. a	8. a	15. e	22. e
2. e	9. a	16. b	23. d
3. e	10. e	17. a	24. a
4. c	11. a	18. d	25. e
5. a	12. a	19. b	
6. c	13. a	20. e	
7. e	14. c	21. e	

What's It All About?

1. **Type of question:** Variation on "defend a position"—asks you to apply your knowledge to describe an observation.

2. **Collect the evidence:**
   What is the relationship between animals and angiosperms? What advantages do animals provide to angiosperms, and vice versa? What disadvantages exist for both organisms? What drives the evolution of organisms?

3. **Pull it all together:**
   The relationship between animals and angiosperms could have evolved only if it provided a selective advantage for survival for both organisms. Producing flowers and fruit requires a lot of energy, but this disadvantage is offset by the advantage that animals are more precise pollinators and more effective at spreading seeds than the wind or water. Animals that adapt to eating nectar or fruit gain the advantage of a high-energy food source that doesn't run away, which usually outweighs the disadvantage of relying on a food source that can change in quality or quantity because of the climate. Because many species of angiosperms and animals exist, we can conclude that the relationship between these two groups is evolutionarily advantageous.

Chapter 23 Animals: The Diversity of Life 3

Compare and Contrast

roundworms/flatworms

Both animals have bilateral symmetry and organs. However, flatworms lack a central body cavity, whereas roundworms have a pseudocoelom, a primitive body cavity.

ectothermy/endothermy

Both are ways of regulating internal temperature. However, in ectothermy the ambient temperature determines an animal's internal temperature, whereas in endothermy the internal temperature is regulated by the animal's metabolism.

triploblastic/diploblastic

Both terms describe the blastula. However, a triploblastic blastula has three germ layers—the endoderm, mesoderm, and ectoderm—whereas a diploblastic blastula has only two germ layers, the endoderm and ectoderm.

bilateral/radial symmetry

Both terms describe symmetry. However, bilateral symmetry refers to an organism in which opposite sides of the body are mirror images, whereas radial symmetry refers to an organism in which body parts are distributed evenly around a central axis, much like a star.

oviparous/viviparous

Both terms describe where fertilized eggs develop. However, oviparous refers to fertilized eggs that are laid outside the body and develop outside the body while viviparous refers to fertilized eggs that develop inside the mother's body.

Short Answer

1. No, the difference is developmental; protostomes have a ventral nerve cord, whereas deuterostomes have a dorsal one.

2. Sponges have microscopic pores between the cells to allow water to wash in nutrients and wash out wastes.

3. Both go through a polyp stage during development.

4. In protostomes, organs develop from mesoderm tissue in the embryo; this tissue is lacking in the Cnidaria and Porifera.

5. Flatworms need to have all their cells within reach of air and nutrients.

6. As they grow, arthropods must molt; a new shell grows beneath the old one, which must be discarded when it gets too small.

7. Insects coevolved with plants to adapt to feeding on angiosperms.

8. Jaws improved the animals' ability to capture and consume prey.

9. The amniotic egg protects the embryo from drying out, decreasing its dependence on a wet environment for survival.

10. Mammals need to consume more food to maintain a high enough metabolic rate.

11. Cephalochordata and Urochordata lack backbones.

Multiple Choice

1. d	8. b	15. c	22. b
2. c	9. d	16. a	23. c
3. e	10. e	17. d	24. c
4. a	11. b	18. a	25. e
5. a	12. a	19. a	
6. c	13. b	20. c	
7. b	14. e	21. a	

What's It All About?

1. **Type of question:** Variation on "defend a position"—asks you to apply your knowledge to describe an observation.

2. **Collect the evidence:** What is the hierarchy of life? (*Hint:* Begin with atoms…) What does this hierarchy describe? What does phylogenetics describe?

3. **Pull it all together:**
Phylogenetics reflects a hierarchy of functions that result from the development of a hierarchy of structures. Classification in Kingdom Animalia reflects gains in structures that support new functions; for example, members of Phylum Porifera are little more than tubes of cells (tissues) that can absorb nutrients from water. By gaining body projections (organs—in this case, tentacles), members of Phylum Cnidaria gain the function of prey capture. Subsequent additions of bilateral symmetry, a body cavity, and so forth, each allow new functions for the organism. Just as the structure of a molecule directs its function, so do the structures found within organisms direct their expanding array of functions.

CHAPTER 24 The Angiosperms: An Introduction to Flowering Plants

Compare and Contrast

pollen/endosperm

Both are associated with angiosperm reproduction. However, pollen contains the male gametes, and the triploid endosperm is contained within the seed and serves as a nutrient source for the embryo.

monocot/dicot

Both are types of angiosperms. However, monocots have one embryonic leaf, whereas dicots have two embryonic leaves.

primary/secondary growth

Both are types of plant growth. However, secondary growth occurs in woody plants and involves an increase in girth, whereas primary growth occurs in all plants and is responsible for increases in height of the plant.

cork cambium/vascular cambium

Both are meristems involved in secondary growth. However, cork cambium produces the outer layer of woody plants, whereas the vascular cambium produces secondary xylem and phloem.

petal/petiole

Both are structures in a flowering plant. However, the petal is part of the flower, whereas the petiole is part of the leaf system.

Short Answer

1. "An apple a day keeps the doctor away" is the common expression. Because the fruit is derived from an ovary, the expression has been reworded to reflect that.
2. Meristem is the embryonic tissue that gives rise to new structures in the plant.
3. The evaporation of water from the leaf creates a flow of water up through the xylem.
4. Thigmotropism allows a plant to use another structure to climb high enough to maximize its exposure to sunlight.

Multiple Choice

1. b	8. c	15. b	22. b
2. b	9. c	16. b	23. c
3. c	10. c	17. a	24. e
4. e	11. d	18. d	25. c
5. d	12. c	19. c	
6. d	13. e	20. c	
7. c	14. d	21. a	

What's It All About?

1. **Type of question:** Variation on "defend a position"—asks you to apply your knowledge to explain an observation.

2. **Collect the evidence:**
   Water transport: Water is pulled into the root to replace water lost by the leaves during transpiration. If more water surrounds the root than transpiration can remove, an aqueous environment that favors bacterial growth surrounds the cells.
   Food transport: A high water concentration in the root cell would dilute the sugar flowing in from the phloem and decrease the osmotic pressure, so less water flows back into the xylem. Water will move up the xylem only if the rate of transpiration is greater than the rate of absorption, so slower osmosis of water out of the roots would slow sap flow. All of this might affect the energy balance of the plant, making it more vulnerable to infection.

3. **Pull it all together:**
   You should see the pattern now—define the changes, and describe how these changes could have come to be.

Chapter 25 The Angiosperms: Form and Function in Flowering Plants

Compare and Contrast

pollen/endosperm

Both pollen and endosperm are essential to reproduction in flowering plants. However, pollen is an immature male gametophyte plant, whereas the endosperm, which is the food for the developing embryo, develops only after the pollen fertilizes the egg.

monocot/dicot

Both terms refer to a way of categorizing angiosperms. However, a monocot has only one embryonic leaf within the seed, whereas a dicot has two embryonic leaves within the seed.

primary growth/secondary growth

Both terms describe patterns of growth. However, primary growth occurs at the tips of shoots and roots, whereas secondary growth refers to lateral growth that thickens the stem or trunk.

cork cambium/vascular cambium

Both terms refer to a type of meristematic tissue. Cork cambium gives rise to the outer tissues of the plant, whereas vascular cambium gives rise to secondary xylem and phloem.

epicotyl/hypocotyl

Both terms refer to parts of a seedling. The epicotyl is that part that will eventually form the leaves, whereas the hypocotyl will form the root system.

Short Answer

1. One way to classify plants is based on the length of the life cycle; another is based on the number of embryonic leaves. A home gardener would be more likely to use the length of the life cycle (annual, perennial, and so on).

2. Parenchyma cells are the basic starting cells from which other cells differentiate and is most analogous to an animal stem cell. This could be demonstrated by growing parenchyma cells in culture and exposing them to plant hormones to cause differentiation.

3. A cross section of a maple leaf shows an upper and a lower epidermis. Underneath the epidermis is a layer of ground tissue that contains photosynthetic parenchyma cells. Adjacent to the actively photosynthesizing cells is the vascular tissue, the xylem and phloem.

4. Yes, it will continue to grow because a lateral bud will become activated.

5. This removes the living phloem, thereby preventing food transport.

6. Water is pulled by the force of evaporation from the leaves.

7. Before fertilization occurs, the pollen grain must land on the stigma of a plant with an ovary. Once this occurs, a pollen tube grows, and the sperm travel down to the egg. The fusion of the haploid sperm and egg form a diploid zygote.

8. A seed is an embryo with a food source in a protected case; fruit is the tissue surrounding the seeds, which develops from the ovary; and pollen contains male gametes.

Multiple Choice

1. e	8. c	15. e	22. d
2. b	9. c	16. a	23. e
3. c	10. c	17. d	24. c
4. c	11. b	18. d	25. a
5. a	12. a	19. c	
6. c	13. d	20. e	
7. c	14. d	21. d	

What's It All About?

1. **Type of question:** Variation on "defend a position"—asks you to apply your knowledge to describe an observation.

2. **Collect the evidence:** What do flowering plants provide that is critical to survival? (*Hint:* Check Sections 25.5, 25.6, 25.7, and 25.8 and Chapter 8 [photosynthesis].)

3. **Pull it all together:**

   After you have listed what you consider to be the five most important functions of plants (and there are certainly more than five), elaborate on the function.

   As an example, plants are the producers of any ecosystem. They produce carbohydrates through photosynthesis, and all living organisms (including plants themselves) rely on carbohydrates as their main energy source.

CHAPTER 26 Introduction to Human Anatomy and Physiology: The Integumentary, Skeletal, and Muscular Systems

Compare and Contrast

cardiac/skeletal muscle

Both are types of muscle. However, cardiac muscle is confined to the heart, whereas skeletal muscle occurs throughout the body.

exocrine/endocrine glands

Both release material that helps to maintain homeostasis. However, exocrine glands release material to the outside (which in this context includes the digestive tract), whereas endocrine glands confine their products to the interior.

basement membrane/ground substance

Both are extracellular structures. However, ground substance is the extracellular material of connective tissues, whereas basement membranes support epithelial tissues.

osteoclast/osteoblast

Both are bone cells. However, osteoclasts break down bone, whereas osteoblasts form new bone tissue.

dermis/epidermis

Both are part of the outer layer of the skin. However, the epidermis is the outermost layer.

Short Answer

1. Connective tissues are categorized by their extracellular material or matrix.
2. Cardiac muscle is found in the heart, skeletal (or striated) muscle moves the body through space, and smooth muscle moves material within the body.
3. Exocrine glands release materials through ducts. Examples are the sweat and sebaceous glands of the skin.
4. Red bone marrow generates blood and immune cells; yellow bone marrow stores fat.
5. Every organ contains different types of tissues.
6. In both cases, a signal stimulates a process that ends up removing the signal. When the temperature drops in the house, the thermostat turns on the heater, which raises the temperature.
7. Ligaments connect bones, and tendons connect muscle to bone.
8. Sebaceous glands produce a waxy secretion that inhibits bacterial growth and lubricates the hair shaft. Sweat glands produce sweat.
9. Homeostasis is the maintenance of the internal environment. Homeostasis is the ultimate motivating factor in all physiological mechanisms.

Multiple Choice

1. d	10. a	19. e	28. a
2. d	11. c	20. b	29. b
3. c	12. d	21. c	30. b
4. a	13. c	22. b	31. e
5. d	14. d	23. e	32. a
6. d	15. a	24. a	33. g
7. b	16. b	25. a	34. f
8. c	17. c	26. c	35. c
9. d	18. a	27. d	36. d

What's It All About?

1. **Type of question:** Two compare-and-contrast questions, which are connected because the goal is to determine whether similar types of tissues have similar types of functions in plants and animals.

2. **Collect the evidence:** Make a table! Unique functions in each organism are in italics.

Plants	Animals
Dermal tissue (epidermis) Functions: protection, gas exchange, *water absorption*	Epithelial tissue Functions: protection, exchange w/environment, *exocrine and endocrine glands secreting hormones*
Ground tissue Functions: forms bulk of plant; structure, storage, *photosynthesis*	Fibrous connective tissue, muscle tissue, loose connective tissue, adipose tissue Functions: forms bulk of animal; provides structure, storage, *packing material*
Vascular tissue (xylem and phloem) Functions: water and food flow	Epithelial tissue Functions: makes the vascular vessels; fluid connective tissue transports water and food; *nervous tissue*

CHAPTER 27   Communication and Control: The Nervous and Endocrine Systems

Compare and Contrast

nerve/neuron

Both are integral components of the nervous system. However, a nerve is a bundle of axons, whereas a neuron is a single nervous-system cell.

steroid hormone/amino acid-based hormone

Both are involved in endocrine signaling and cause responses in target cells. However, amino acid-based hormones bind to receptors on cell surfaces, whereas steroid hormones enter cells to exert their effects.

pituitary/hypothalamus

Both are nervous system structures involved in endocrine control. However, the pituitary controls the release of hormones.

axon/dendrite

Both are part of a neuron. However, the dendrite receives information, and the axon transmits information from the cell body to the synaptic terminal.

sympathetic/parasympathetic

Both are parts of the autonomic nervous system. The sympathetic division has stimulatory effects on the body, whereas the parasympathetic division has relaxing effects on the body.

Short Answer

1. In the nervous system, communication occurs by means of cell-to-cell contact; the endocrine system uses the circulatory system to carry messages.

2. Rods and cones are photoreceptor cells found in the eye. Rods respond primarily to low-light situations, whereas cones respond to bright light and, therefore, color.

3. Homeostasis is maintained primarily by negative feedback.

4. Hormones are secreted by endocrine glands, which are unique in that they have no ducts.

5. Steroid hormones are derived from cholesterol. A certain amount of cholesterol is necessary to make these essential hormones (which include the sex hormones).

6. Hormones alter cellular activities by changing the production or shape of enzymes or structural proteins.

7. Because negative feedback removes the stimulus instead of increasing it. This is important if the stimulus is something like low blood sugar.

8. The motor neurons are responsible for turning the head; the sensory neurons, for smelling the rose; and the interneurons and motor neurons, for picking up the rose.

9. Ions flow in when the sodium channels open and then flow out when the potassium channels open. The membrane potential changes in one section of the membrane, whereupon the ion gates are triggered to open in the next section of the axon membrane.

10. Neurotransmitters communicate between neurons at synapses. They do not all send the same message; for example, acetylcholine is stimulatory, whereas dopamine is inhibitory.

11. The pulling back is due to the reflex arc running from the sensory neuron in your hand through the spinal cord to a muscle, which contracts in response.

12. Hypoglycemia can occur in diabetics who take medications to keep their blood sugar under control and may be a result of not eating enough food or of increased activity. The problem can be corrected by taking a *small* amount of sugar. However, too much sugar is not good and may cause an imbalance in insulin levels. If you are a diabetic, you should check your blood glucose at regular intervals and adjust your insulin dosage accordingly.

13. Damage to hair cells can cause the roots of cilia to break off or become permanently damaged; brain injury may affect the signals that the brain receives from the hair cells wtihin the basilar membrane; a torn eardrum will interfere with the initial vibration as sound waves enter the ear canal; buildup of earwax can dampen the vibration of the three bones of the middle ear.

14. Both senses rely on collections of nerve endings that are receptors. In taste, the receptors are located on small bumps in the tongue called papillae. Contact with a particular taste causes release of a neurotransmitter. In touch, there are special receptors in the skin that can distinguish five different types of touch. Contact with a specific receptor causes a signal to be sent to the brain.

15. A person can survive in a vegetative state provided that the brain stem is intact because the medulla oblongata controls the basic functions critical to life, such as breathing, heart activity, and digestion.

16. The pituitary is referred to as the master gland because it has two sections, both of which produce and release specific hormones. However, the release of these hormones is controlled by the hypothalamus, by the section of releasing hormones that cause the pituitary cells to produce their own hormones.

Multiple Choice

1. b	8. c	15. e	22. a
2. c	9. d	16. d	23. a
3. e	10. d	17. a	24. a
4. d	11. c	18. b	25. d
5. b	12. d	19. b	
6. d	13. c	20. e	
7. d	14. c	21. b	

What's It All About?

1. **Type of question:** Analysis—asks you to think about how we see and try to figure out how being blind for a long time affects the system.

2. **Collect the evidence:**
   a. All sensory neurons deliver information to the brain.
   b. Visual system: detects light, focuses image; converts image into neural signal; brain integrates neural signal.

3. What does it mean to integrate a signal? To combine the signal with information from other senses and experience. Experience is knowing that an item shaded on the bottom is raised above the surface—if you've never learned that, will it make sense the first time you see it?

CHAPTER 28 Defending the Body: The Immune System

Compare and Contrast

specific defense/nonspecific defense

Both are mechanisms of the immune system. However, the specific defenses result in the formation of antibodies and immunity while the non-specific do not.

B cell/T cell

Both are part of the specific defense. However, while B cells generate the formation antibodies, T cells are part of the cell-mediated defense.

histamine/cytokine

Both are molecules involved in the immune system. However, histamine is involved directly in the inflammatory response while cytokine plays roles in both specific and non-specific defenses.

dendritic cells/macrophages

Both are antigen presenting cells. However, macrophages engulf pathogens and cellular debris while dendritic cells do not.

plasma cells/B cells

Both are part of the specific immune response. However, B cells generate plasma cells which in turn produce antibodies.

Short Answer

1. When the immune system attacks normal body tissue that is described as an autoimmune disorder. An example would be rheumatoid arthritis where the immune system has attacked the cartilage lining the joints of the skeletal system.

2. A vaccine is comprised of antigens for a specific pathogen. By inoculating the body with that portion of the disease-causing organism, the immune system is stimulated to produce antibodies and therefore, create immunity.

3. The inflammatory response is a non-specific response to infection or trauma. The signs of the inflammatory response are redness, heat, swelling, and pain at the site of injury. Inflammatory responses may also be systemic as well.

4. Both are infectious diseases caused by viruses. However, their target cells are different. Polio attacks the nervous system while HIV (the causative agent of AIDS) attacks the helper T cells of the immune system.

5. Antigen presenting cells alert the specific immune system to the presence and type of antigen (and therefore pathogen) present in the body.

6. Helper T cells: oversee the specific immune response by activating B cells and cytotoxic T cells. Cytotoxic T cells: generate the cellular immune response that destroys pathogen-infected body cells. Regulatory T cells: inhibit or decrease the specific immune response so as to prevent a runaway response to the pathogen.

7. Antibodies bind to their antigens and thereby make them targets for phagocytosis. They also prevent the pathogens attached to those antigens from infecting the body cells, thereby increasing the rate at which the immune system can rid the body of infection.

Multiple Choice

1. b	8. e	15. d	22. b
2. c	9. b	16. e	23. c
3. d	10. d	17. c	24. a
4. d	11. e	18. e	25. a
5. d	12. b	19. c	
6. d	13. a	20. c	
7. a	14. b	21. e	

What's It All About?

1. **Type of question:** Defend a position, and the question could go either way.

2. **Collect the evidence:**
   a. Specific immune system depends on exposure to a non-self molecule (antigen).
   b. Antigen binding to B cells elicits antibodies; antigen binding to macrophages activates T cells.
   c. In both cases, the B or T cell has to be able to recognize the antigen, meaning that it must have the receptor.

   On the one hand, early exposure to antigens shouldn't matter, because if you have circulating B cells that recognize a self-protein, your body can trigger an autoimmune reaction no matter how many other antigens it sees.

   But on the other hand, maybe early antigen exposure does protect you, because exposure to a large number of antigens creates large numbers of circulating memory cells. As a result, the percentage of B cells sensitive to self antigens are diluted in the B cell population, making it very unlikely that a B cell carrying a self protein receptor will ever locate its antigen.

CHAPTER 29 Transport and Exchange 1: Blood and Breath

Compare and Contrast

red blood cells/white blood cells

Both are formed elements of blood. However, RBCs carry oxygen, using hemoglobin, whereas WBCs fight foreign invaders.

artery/vein

Both carry blood in the circulatory system. However, arteries carry blood away from the heart, whereas veins carry blood toward the heart and have valves.

high-density lipoprotein/low-density lipoprotein

Both are types of proteins that transport lipids. However, high-density lipoproteins are known as the "good" cholesterol because they transport cholesterol from various parts of the body to the liver, whereas low-density lipoproteins are known as the "bad" cholesterol because they transport cholesterol from the liver to various tissues.

respiration/ventilation

Both terms refer to oxygen delivery. However, respiration is the physical inhalation of gases, whereas ventilation is the physical movement of air into and out of the lungs.

alveoli/bronchiole

Both are parts of the respiratory system. However, the alveoli are the actual sites of gas exchange, whereas the bronchioles are the fine tubes that deliver oxygen to the alveoli.

Short Answer

1. A capillary is single-cell-thick with an epithelium lining; an artery has three layers; and a vein has three layers with valves.

2. Two chambers receive blood into the heart, and two pump blood out of the heart.

    Two handle oxygenated blood, and two handle unoxygenated blood.
    (i) Right atrium receives unoxygenated blood from the body.
    (ii) Right ventricle pumps unoxygenated blood to the lungs.
    (iii) Left atrium receives oxygenated blood from the lungs.
    (iv) Left ventricle pumps oxygenated blood to the body.

3. The systemic circulation drops off oxygenated blood at the cells of the body, whereas the pulmonary circulation picks up oxygen from the lungs.

4. Coronary arteries are blocked in a heart attack—they normally supply blood and oxygen to the heart muscle.

5. Exchange of gases occurs in the capillaries. The lymphatic system assists in returning blood back toward the heart, along with skeletal muscles that help to propel blood back toward the heart.

6. Osmosis is important in the exchange of nutrients because there is a concentration gradient difference between solutes in the blood and solutes in the cells. Thus, water leaves the capillaries at the arterial end (where blood is hypotonic) and reenters the capillaries at the venous end (where blood is hypertonic).

7. Humans inhale air as the rib cage and muscular contractions of the diaphragm expand the chest.

8. The surface area of the alveoli must be large to provide sufficient surface area to exchange enough oxygen for the needs of every cell in the body.

Multiple Choice

1. d	8. c	15. b	22. c
2. c	9. a	16. a	23. c
3. d	10. a	17. c	24. b
4. c	11. e	18. b	25. b
5. d	12. a	19. a	
6. c	13. d	20. c	
7. e	14. e	21. b	

Answers to heart diagram:

1. right atrium
2. right ventricle
3. pulmonary arteries
4. pulmonary veins
5. left atrium
6. left ventricle
7. aorta

What's It All About?

1. **Type of question:** Apply your knowledge: what is the function of the circulatory system (*Hint:* What is the title of this chapter)?

2. **Collect the evidence:** Role of circulation is to carry to tissues and waste products away.
    a. Fatigue—indicates lack of O_2 to the muscles
    b. Lethargy—indicates lack of O_2 to the brain
    c. Infections—indicates lack of O_2 to immune cells; less cellular function because energy is not made (remember respiration, discussed in Chapter 7).
    d. Cold feet—we lose energy as heat during metabolism, which makes our blood warm. If the blood doesn't get to the toes, they stay cold.

CHAPTER 30 Transport and Exchange 2: Digestion, Nutrition, and Elimination

Compare and Contrast

fat/fibers

Both are part of a healthy diet. However, whereas fat provides energy to fuel metabolism, fiber is mostly indigestible.

vitamin/mineral

Both are essential parts of the diet. However, vitamins are necessary as coenzymes.

mechanical digestion/chemical digestion

Both are part of the breakdown of food. However, mechanical digestion involves the teeth and various muscles, whereas chemical digestion relies on enzymes and acids.

ureter/urethra

Both are structures of the urinary system. However, ureters carry urine to the bladder from the kidney, whereas the urethra carries urine from the bladder to the outside.

small intestine/large intestine

Both are part of the digestive system. However, the small intestine is involved primarily in the absorption of nutrients, whereas the large intestine is involved in the formation of feces.

Short Answer

1. See Figure 30.2.
2. Urine is formed in the kidneys. The urine is then carried by the ureters to the bladder. Urine is subsequently released from the bladder through the urethra.
3. The liver forms bile, which is necessary for fat digestion.
4. Secretion is the active removal of material from the bloodstream. Examples of secreted materials include penicillin and some ions.
5. The urethra is much longer in males, a characteristic that makes it more difficult for bacteria to cause infection.
6. Most of the water entering the digestive system each day is absorbed in the small intestine. Much of the remainder is absorbed by the large intestine.
7. The kidney is able maintain blood volume by reabsorbing almost all of the water that enters the kidney during filtration, and by selectively moving ions, as well.
8. Digestion of polymers and absorption of nutrients.
9. The six classes of nutrients are as follows: (1) proteins, which provide amino acids for structural and metabolic functions; (2) lipids, which provide energy and the material for building lipid structures; (3) carbohydrates, which provide energy; (4) water, which helps to maintain blood volume and cell volume; (4) vitamins, which function as coenzymes; and (6) minerals, which provide material for body structure and metabolic function.
10. The stomach stores food and allows the mechanical and chemical breakdown of food.
11. The liver and gallbladder produce and store bile, which is needed for lipid digestion. The pancreas produces enzymes and bicarbonate to complete digestion in the small intestine.
12. A food pyramid is a dietary aid that lists the different types of nutrients by their relative amounts in a healthy diet.
13. The colon and its bacteria have a symbiotic relationship. The bacteria need the nutrients in the colon and in turn produce vitamins that the body requires.
14. A nephron is the functional unit of the kidney. It is there that the vast majority of water and salt is reabsorbed and returned to blood.

Multiple Choice

1. c	8. c	15. d	22. d
2. b	9. e	16. d	23. e
3. b	10. a	17. a	24. d
4. e	11. d	18. b	25. a
5. a	12. a	19. b	
6. c	13. b	20. d	
7. d	14. b	21. a	

What's It All About?

1. **Type of question:** Apply your knowledge to answer a question.

2. **Collect the evidence:** How are proteins, carbohydrates, and lipids digested and absorbed in the body?

3. **Pull it all together:**
   The cheeseburger is mostly protein and fat. The protein begins to be digested in the stomach, a process that continues in the small intestine. The fat is digested in the small intestine. The simple carbohydrates in the bun begin to be digested in the mouth and are completely digested in the small intestine.

CHAPTER 31 An Amazingly Detailed Script: Animal Development

Compare and Contrast

mesoderm/ectoderm

Both are germ layers of the gastrula. However, the mesoderm is the central layer, whereas the ectoderm is the outer layer. Each layer develops into different final tissues in the adult organism.

notochord/neural tube

Both are specific structures formed during early organogenesis. However, the notochord is a flexible rod that induces the formation of the neural tube from ectoderm, whereas the neural tube ultimately forms the CNS.

blastula/gastrula

Both are intermediate structures formed during early development. However, the blastula is a hollow ball of cells, whereas the gastrula is a three-layered embryo that develops from the blastula.

determined cells/committed cells

Both refer to developmental stages of a cell with respect to its ultimate fate. However, determined cells have a recognizable fate (which may be changed under the influence of morphogens), whereas committed cells have a fate that cannot be reversed.

morphogen/transcription factor

Both are chemicals that control development. However, a morphogen is a substance whose concentration affects development in a specific embryonic region, whereas a transcription factor is a substance that directly affects the production of a specific protein by binding to DNA.

Short Answer

1. Development continues as a result of the continuing interactions between genes, proteins, and the environment.

2. Induction is the capacity of some embryonic cells to control the development of others.

3. Morphogens are substances that can diffuse through the embryo, influencing development by their relative concentration.

4. Homeobox genes were discovered in *Drosophila*. They relate to development by producing transcription factors, which, in turn, control the synthesis of specific proteins.

5. Three phases of development:

Name of Stage	Description	Function
Cleavage	Rapid division of zygote	Formation of the blastula into many cells
Gastrulation	Cell movements and migrations	Creates the three germ layers
Organogenesis	CNS forms by rolling of a tube of ectoderm	Formation of specialized organs

Multiple Choice

1. e	8. c	15. c	22. c
2. b	9. d	16. b	23. c
3. c	10. d	17. e	24. a
4. b	11. c	18. b	25. a
5. e	12. a	19. b	
6. d	13. d	20. b	
7. d	14. a	21. c	

What's It All About?

1. **Type of question:** Apply your knowledge to answer a question.

2. **Collect the evidence:**
   Phylogeny
   - Simple organisms are ancestral to complex ones.
   - Complexity involves gaining specialized structures for new functions.
   - Controlled by genes (mutations are fuel for natural selection)
   Ontogeny
   - Early stages produce general structures.
   - Later stages produce more complex structures, with specific functions, from the simpler structures.
   - Directed by gene products

CHAPTER 32 How the Baby Came to Be: Human Reproduction

Compare and Contrast

sperm/semen

Both are components of the male ejaculate. However, sperm are the male gametes, and semen is the fluid containing sperm and other compounds (nutrients and lubrication).

uterine tube/urethra

Both are structures associated with the human genitourinary tract. However, uterine tubes are found only in females. Both males and females have a urethra, which discharges urine (and sperm in males).

epididymis/corpus luteum

Both are structures in the human reproductive anatomy. However, the epididymis is a site of sperm storage within the testes, whereas the corpus luteum is the structure remaining after the ovulation of an oocyte from the follicle.

oocyte/zygote

Both are cells involved in development. However, an oocyte is the female gamete, whereas a zygote is the fertilized (and therefore diploid) egg.

inner cell mass/trophoblast

Both are structures that form early in embryonic development. However, the trophoblast will become part of the placenta, whereas the inner cell mass forms the embryo.

Short Answer

1. The three stages of birth are cervical dilation, expulsion of the baby, and expulsion of the placenta (afterbirth).

2. Internal body temperatures are too high for viable sperm production.

3. Identical twins result when one sperm fertilizes one egg. Fraternal twins result when two sperm fertilize two eggs.

4. Prior to 28 weeks of gestation, surfactant is not produced, so the lung tissue sticks to itself when premature babies try to breathe.

5. The term *fetus* is used after week 9 post-fertilization.

6. A follicle is the egg and the support structures that surround it.

Multiple Choice

1. c	8. c	15. a	22. e
2. c	9. b	16. a	23. d
3. a	10. d	17. d	24. a
4. b	11. c	18. e	25. d
5. d	12. b	19. a	
6. c	13. c	20. c	
7. d	14. b	21. b	

What's It All About?

1. **Type of question:** Apply your knowledge to answer a question.

2. **Collect the evidence:**
   The only evidence we have is the observation that the embryo is not attacked by the maternal immune system (in most cases). Scientists are only beginning to approach this question experimentally as we learn about the immune system in greater detail. We can't answer this question directly, but we can propose some testable hypotheses to explain this observation, just as a practicing scientist might.

   Hypothesis 1: The embryo produces something that blocks the maternal immune system from recognizing non-self antigen, so we might look for novel chemicals in fetal blood.

   Hypothesis 2: The embryo produces something that inactivates the maternal immune system, so we would look for an inhibitor of immune function. If this were happening, we might expect that pregnant women would get sick more often than nonpregnant ones.

   Hypothesis 3: The maternal immune system shuts down part of its immune system, such as B-cell surveillance, to protect the embryo, so we might find that pregnant women have fewer circulating B cells than nonpregnant women.

   Hypothesis 4: The embryonic blood never directly contacts maternal blood. Immune cells are too large to diffuse through the membranes, so the maternal immune system never sees antigens from the embryo. In this case, we should look for immune cells in the "wrong" place (fetal cells in maternal circulation).

CHAPTER 33 An Interactive Living World 1: Populations in Ecology

Compare and Contrast

density dependent/density independent

Both are factors that affect population growth. However, density-dependent factors usually involve biological factors, while density-independent factors usually involve physical factors.

r-selected species/K-selected species

Both refer to ways of categorizing species within a population. However, an *r*-selected species tends to fluctuate widely in response to environmental conditions, whereas a *K*-selected species tends to have a rather stable population, at or near the carrying capacity of the ecosystem.

arithmetical increase/exponential increase

Both refer to patterns of population growth. However, a population that is arithmetically increasing grows by a fixed number at each population interval, whereas a population showing an exponential growth increases by a number proportional to the number of individuals in the population.

community/population

Both refer to ways in which biological organisms interact. However, populations are collections of interacting individuals, and communities are collections of interacting populations.

exponential growth/logistic growth

Both refer to shapes of population growth curves. A population with exponential growth will show a J-shaped curve, whereas a population with logistic growth will show an S-shaped curve.

Short Answer

1. Please refer to the text for a pre-reproductive age population (Kenya) and for a post-reproductive age population (United States).

2. The three parameters involved in population dynamics are count (number of individuals); distribution of individuals over a geographical area; and changes in the size of a population over time.

3. Immigration and emigration also influence population change.

4. Label curves: Please refer to the text.

   exponential   fruit flies
   arithmetic   number of biology textbooks leaving the printing press

Multiple Choice

1. d	6. d	11. d	16. e	21. c
2. c	7. d	12. a	17. e	22. b
3. a	8. c	13. b	18. c	23. d
4. c	9. c	14. b	19. b	24. b
5. a	10. d	15. c	20. a	25. a

What's It All About?

1. **Type of question:** Apply your knowledge to answer a question.

2. **Collect the evidence:**
   - Identify the type of population growth that the deer are showing. What are possible limiting factors on the population, both biotic and abiotic?
   - How do the deer interact with other populations within the community?
   - How will you be able to explain to citizens the dynamics of the deer population?
   - Once you find answers to these questions, begin to develop several solutions to the problems, and begin testing!

CHAPTER 34 An Interactive Living World 2: Communities in Ecology

Compare and Contrast

competition/predation

Both are ways in which populations can interact. However, competition occurs when individuals attempt to use the same resources, whereas predation is the consumption of one species by another.

primary/secondary succession

Both refer to ways in which communities change over time. However, primary succession is the establishment of an initial community in an area with no previous living organisms, whereas secondary succession is the change of the community as it becomes established.

community/population

Both refer to ways in which biological organisms interact. However, populations are collections of interacting individuals, and communities are collections of interacting populations.

mutualism/commensalisms

Both refer to specific interactions between individuals. However, mutualism is a situation in which both individuals benefit, whereas in commensalism one individual benefits and the other is unaffected.

coexistence/coevolution

Both refer to shared interaction. However, coevolution refers to a specific series of adaptations caused by the interactions.

Short Answer

1. This description best fits commensalism.

2. The common processes are facilitation of growth by earlier species for later ones and competition among later species.

3. Parasitism is a variety of predation in which the predator feeds on prey but is unlikely to kill it.

Multiple Choice

1. d	8. e	15. e	22. c
2. d	9. c	16. d	23. d
3. d	10. c	17. b	24. b
4. c	11. c	18. e	25. b
5. c	12. d	19. b	
6. b	13. e	20. d	
7. a	14. e	21. b	

What's It All About?

1. **Type of question:** Defend a position. Given what you know about viruses and the human body, can this relationship be described as parasitic?

2. **Collect the evidence:** Make a table!

Yes, it is parasitic	No, it is not parasitic
• Viruses benefit; hosts do not. • Viruses compromise host's functioning by sucking up resources. • Viruses can kill their host. • Viruses can evolve to avoid the host's immune system.	• Hosts do benefit by developing antibodies and T cells against that virus, providing long-term protection. • Viruses do not infect hosts for nourishment; viruses aren't alive. • By weakening or killing the host, the virus would fail to guarantee its own survival.

CHAPTER 35 An Interactive Living World 3: Ecosystems and Biomes

Compare and Contrast

herbivore/detritivore

Both are consumers in a food chain or food web. However, herbivores eat producers (plants), whereas detritivores consume dead organic material, serving as the recyclers.

gross primary production/net primary production

Both refer to conversion of the sun's energy to plant biomass by photosynthesis. However, gross primary production is the total amount of energy assimilated, whereas net primary production is the total amount of energy assimilated, less the amount of energy required to maintain the plant.

estuary/wetland

Both are bodies of water. However, estuaries are the areas where streams or rivers flow into the ocean, whereas wetlands are wet (submerged) for a portion of the year.

ecosystem/biome

Both refer to ecological levels of organization. However, ecosystems include the organisms and their biotic and abiotic environments, whereas a biome is a community defined by the vegetation.

abiotic/biotic

Both have an effect on ecosystems. However, abiotic factors are nonliving components, and biotic factors include all living components of an ecosystem.

Short Answer

1. The two types of factors that help to mold ecosystems are conditions and resources.

2. A community consists of various populations living together in the same area at the same time. When you add the abiotic factors, it is referred to as an ecosystem.

3. Increased carbon dioxide comes from burning of fossil fuels plus intensive burning of forested areas.

4. Through the process of decomposition by soil bacteria and fungi

5. **a.** Starting material—atmospheric nitrogen; end product—ammonia

   b. It is the only natural source of nitrogen that is usable by organisms.

6. The source is sunlight, which hits the equator with the most intensity and then sets up the global convection of air currents.

7. More carbon is available to plants in the atmosphere through the burning of wood, coal, and fossil fuels.

8. Answers to diagram: 1st: sunflower; 2nd: blue jay ; 3rd: house cat; 4th: coyote

Multiple Choice

1. d	8. c	15. d	22. c
2. a	9. d	16. a	23. d
3. e	10. b	17. a	24. b
4. b	11. c	18. c	25. e
5. c	12. a	19. d	
6. d	13. b	20. a	
7. b	14. c	21. d	

What's It All About?

1. **Type of question:** Apply your knowledge to identify essential components of a self-contained Earth environment.

2. **Collect the evidence:** Identify and describe the challenges. You might want to do some Internet searching and learn about Biosphere 2, a 3.1-acre, glass-enclosed facility that was the site of two experiments with humans living in a self-contained environment.

   Regulation of the carbon cycle—need to provide controlled environment (temperature, humidity, soil nutrients) for the plants so that the amount of carbon dioxide produced by the nonplant organisms is sequestered appropriately. Too much would make people lethargic, and too little would inhibit photosynthesis, compromising the food supply. Also, need to worry about the plants making enough oxygen.

   Regulation of the hydrologic cycle—everything inside the community needs water, but the water vapor doesn't have a near-infinite space to expand into. Excess water collecting on surfaces would favor growth of mold and bacteria—not a favorable human environment. The consequences of lack of water are obvious.

   Nitrogen cycle—dealing with waste and decay is one of the hardest issues to resolve. Nitrogen-fixing plants would need to be included in the mix, but there would also be a concern about excess nitrates in the soil.

CHAPTER 36 Animals and Their Actions: Animal Behavior

Compare and Contrast

navigation/migration

Both are part of long-distance movement. However, navigation is an aspect of migration.

proximate/ultimate causes of behavior

Both cause a given behavior to occur. However, proximate causation involves stimuli occurring right now, whereas ultimate causation is the effect on an individual's fitness in evolutionary terms.

taxis/reflex

Both involve responses to stimuli. However, reflex movements are extremely well defined, whereas taxis movements are more general.

habituation/imprinting

Both involve organisms tailoring their behavior in response to the environment. However, habituation is a decrease in response, whereas imprinting is a refinement.

classical conditioning/operant conditioning

Both are learning mechanisms. However, operant conditioning involves trial-and-error learning, whereas classical conditioning links an unrelated stimulus to an existing reward pathway.

Short Answer

1. Natural selection, as the agent of evolution, is responsible for the ultimate causation of behavior.

2. Whereas external influences represent different aspects of the environment, internal influences focus on genetics and physiological responses.

3. A classic example is that of the greylag goose, which reflexively retrieves round, egg-like objects for her nest.

4. Two examples are cited in the text: egg-fanning behavior in cichlid fish and singing behavior in male birds.

5. Altruism is the performance of a behavior at some cost or risk to the performer; reciprocal altruism is the performance of a behavior with expectation of future return.

6. Costs of being a social species include increased visibility to predators, decreased availability of resources (food, and so on), and increased likelihood of communicable disease. Benefits include cooperative behavior (hunting, offspring care), altruistic interaction, and defense.

7. Male birds have an innate (genetic) component to their song, but in many cases they also tailor their song to their local environment (other males, and so on).

Multiple Choice

1. e	8. b	15. a	22. d
2. a	9. d	16. c	23. a
3. d	10. e	17. b	24. e
4. c	11. d	18. c	25. b
5. c	12. d	19. b	
6. e	13. c	20. b	
7. b	14. d	21. a	

What's It All About?

1. **Type of question:** Tricky.

2. **Collect the evidence:**
Given the frequency of abandonment and child abuse in our modern world, you could argue that my premise, that humans demonstrate altruism, is false. Yet, if you examine the broader human experience, as described in stories, in rituals, in societal values, you could find the evidence to support the contention that humans are altruistic animals, primarily to their kin but also to strangers. Why? Maybe biology can explain life well and behavior reasonably well, but it needs the help of literature, philosophy, ethics, and the arts to explain love. The natural sciences are only one piece of a liberal arts education—remember to apply the arts and humanities too when solving the truly important questions of this world.